The Wingsnappers

The Wingsnappers

Lessons from an Exuberant Tropical Bird

Barney A. Schlinger

Yale UNIVERSITY PRESS

New Haven & London

Published with assistance from the foundation established in memory of
Amasa Stone Mather of the Class of 1907, Yale College.

Yale University Press books may be purchased in quantity for educational,
business, or promotional use. For information, please e-mail
sales.press@yale.edu (U.S. office) or sales@yaleup.co.uk (U.K. office).

Set in Janson Roman type by Integrated Publishing Solutions,
Grand Rapids, Michigan.
Printed in the United States of America.

Library of Congress Control Number: 2022951524
ISBN 978-0-300-26941-3 (hardcover : alk. paper)

A catalogue record for this book is available from the British Library.

This paper meets the requirements of ANSI/NISO Z39.48-1992
(Permanence of Paper).

10 9 8 7 6 5 4 3 2 1

To all of my colleagues, students, and postdoctoral fellows whose work on manakins made this book possible.
And to Lorie.

Contents

Contents

Contents

Color plates follow page 82

Preface

As a high school senior in 1972–1973, I took an honors biology class. The teacher had shoulder-length brownish hair and round wire-rimmed John Lennon glasses. He was unlike all my other instructors, a real East Coast intellectual. What I remember is making (extracting) DNA in the lab portion of the class by swirling a mucus-like mass around in a test tube. At the time, and according to the teacher, this was cutting-edge high school biology.

In my spare time, I read *In the Shadow of Man*, the seminal 1971 work by Jane Goodall on the chimpanzees of Gombe Reserve that she had studied and with whom she had bonded. I was hooked. When required to write a term paper for my biology class, the biggest assignment of the year, I chose to write about animal tool use. To my great disappointment, my teacher gave my lengthy tome a C, saying that "tool use by animals is NOT biology."

Well, I was seventeen, with way too much testosterone activating all my fresh little androgen receptors, and I was furious. In my mind, of course animal tool use was biology, a perspective confirmed when, later that year, Niko Tinbergen, Karl von Frisch, and

Konrad Lorenz shared the Nobel Prize in Physiology/Medicine "for their discoveries concerning organization and elicitation of individual and social behaviour patterns" (to quote the Nobel Prize webpage). They were recognized for their studies of the behavior of wasps, bees, and geese. Clearly, animal behavior was biology, and tool use was behavior.

When I headed to college at Tufts University (the farthest school from Dallas that accepted me), I instantly declared a biology major. Unfortunately, college biology instruction was dictated by medical school entrance requirements that forced students to take multiple semesters of chemistry, physics, and calculus. Added to these courses were the other breadth requirements, leaving little time for "real" biology. My testosterone kicked in again and I quietly rebelled against this structure, deciding simply to take as many biology courses as possible: botany, evolution, animal behavior, and more. I discovered lichens and liverworts and ulva and goldeneyes and scoters and buffleheads. It was then that biology came alive for me.

Jane Goodall's research and those Nobel Prizes for research in animal behavior aside, the early 1970s are best known as an explosive era for molecular biology. This field dominated all biomedical research and most biomedical funding, and to this day largely defines biological inquiry. Yet while the study of nucleotides opened a vast number of doors, it also sucked the life out of organismal biology. When E. O. Wilson (one of my heroes) tried to resuscitate suprareductionist biological thinking in 1975 with his powerhouse *Sociobiology*, he met immense resistance.

In an unlikely way, Wilson actually contributed to another reductionist hit on biological thought. Ecology, that supraorganismal study of life that blossomed as science during the environmental movement of the 1960s, was found, in part by Wilson, to be framed best mathematically. Of course, experimental biology had always been a quantitative enterprise, but mathematical modeling of ecological phenomena now converted ecosystems into equations.

These two forces, molecular and quantitative biology, led to countless profoundly important discoveries over the next half century. No argument here. But what these two forces also did was turn the attention of many life scientists away from the wonder and diversity of individual organisms. Instead, biology became the study of life's numbers and parts.

Each organism is indeed an intricate collection of all of the principles and mechanisms outlined by reductionist thought, but it is also so much more—a unique living package that defies imagination. We readily accept the concept of the mind as an emergent property of the brain. In the same way, an organism's life is an emergent property of its collection of anatomical and physiological specializations. After fifty years of tearing organisms apart, it is time for us to piece them back together, before it's too late for them and maybe for us as well.

This book is focused on one organism, the golden-collared manakin (*Manacus vitellinus*) of the Panamanian rainforests. Males of this species are beautiful and athletic and noisy, and perform an extraordinary courtship display to attract females for mating. The females see and hear and assess each male before mating. All of this is accomplished against the backdrop of heat, clouds, and rain, and the wet, lush, greenest of greens: the tropical rainforest, with its monkeys, jaguars, toucans, parrots, and insects galore—and so much more.

To be a male manakin requires more than just good looks; rather, it involves a myriad of specialized neurons, muscles, bones, and hormonal systems. Instructions for all of this anatomy- and physiology-based behavior are encoded in the unique collection of manakin genes. To understand how these work in concert requires biomechanical thinking as well as an appreciation of the endocrine control of neuronal firing and muscle contraction. Indeed, an entire course in biology exists within a single manakin—and by the same token, the discoveries of many dedicated biologists can be seen at play each and every moment in the life of a male manakin.

This book is an attempt to capture this complicated biology, by telling the holistic story of one organism, a small, exuberant tropical bird.

Some of you may find concepts like biomechanics and physiology and genetics daunting. As you fall in love with this bird, however, you will want to know all about these ideas; they will fascinate you and freshen your perspectives on the science and wonder of biology. Hopefully, you will also be inspired to take a new look at other organisms, including the one we pay attention to most of all: ourselves.

I have attempted to communicate observational and experimental findings within the stories I tell about the birds. Science is, however, driven by data. Those who wish a deeper dive into the science might seek out the original publications included in the Recommended Reading section. There you will find a more critical evaluation of the methodologies involved. I also recommend taking your own trip to the Panamanian rainforest.

PART I

Birds, Manakins, and Biology

ONE

Setting the Panamanian Stage

Roughly midway between Colombia to the southeast and Costa Rica to the west lies an extensive, relatively flat lowland separating two Panamanian mountain chains. This gap between the mountains, running roughly northwest to southeast, was, for centuries, recognized as a narrow fordable passageway between the Atlantic and Pacific oceans. Trails following the great Chagres and Rio Grande rivers constituted a trading route for indigenous peoples. Upon its discovery by Europeans, the gap became the overland route traversed by conquistadors and, ultimately, Spanish traders and settlers. In the early nineteenth century, this narrow link was eyed by major political and economic powers, and in 1849 permission was granted for the Pacific Mail Steamship Company, based in New York, to construct and operate a railroad following those rivers and ancient trails, thereby connecting the Atlantic and Pacific.

In the mid to late 1800s, treasure hunters and brave immigrants from the eastern United States and Europe wanting to reach the advertised prosperity in California had three choices. One was to cross North America and risk fateful encounters with indigenous warriors protecting their homelands, or becoming lost and dying from thirst or starvation. Alternatively, travelers could board a marginally seaworthy ship and sail around South America, including through the dangerous Drake Passage. A third option was to sail to Colón, Panama, then hike, ride a mule, or, eventually, hop the train south to Panama City, a fifty-odd-mile journey, to catch a second ship sailing north to California. This last option was by far the preferred one for many, and became a popular route. Much of the California "gold rush" involved a slow crawl through central Panama.

Of course, a trip through the Panamanian jungle was challenging. Deadly snakes with names like fer-de-lance and bushmaster were common; jaguars and pumas stalked the nighttime forest. During the eight-month rainy season huge amounts of rain fell, often three to four meters or more, washing away trails and leaving the forest floor a soggy mess crisscrossed by innumerable streams. The heat was taxing, and the humid air dense with mosquitos and biting flies, bees and wasps. Among the constant threats were malaria, which, if it did not kill you outright, left you weak and sick, and yellow fever, which melted your organs and finished you off in agony. Dysentery was rampant. Many travelers were robbed along the way.

Setting these worries aside, however, the travelers also experienced the magical and magnificent rainforest. The air was perpetually thick with the pungent smell of life. At night, frog calls emanated from every nook and cranny. During the day, everything everywhere was deep, deep green, embellished here and there with giant trees that flowered profusely in yellows, reds, or purples: acres of blossoms atop the hundred-foot forest canopy. Circling the immense cumulus clouds that grew in the morning heat flew vast flocks of vultures, kites, swifts, and martins. Inside the forest buzzing with sounds of katydids and cicadas, huge blue butterflies leisurely flapped up and down the trails and clearings. Howler monkeys screamed with each clap of thunder. Macaws and parrots shrieked, toucans croaked, and birdsong of all types filled the air.

Periodically, amid this extraordinary cacophony, travelers were surprised to hear rounds of explosive snapping sounds emanating from the surrounding forest. In some cases, whole patches of forest along the trail and train tracks from Colón to Panama City seemed to burst with these raucous snaps.

John Muir made his way from New York to California in 1868 through Panama. As the train roared along at a "cruel speed," Muir moaned and could "only gaze from the car platform & weep."[1] He had no time for botanical exploration, because he had to catch a ship in Panama City. His trip occurred during the peak of the Panama dry season, when the forest snapping sounds are at their loudest. Undoubtedly, despite the rattling of the train cars and the constant thumping of the engine, Muir heard all this noise and wondered what on earth it could be.

At that time, Panama was part of Colombia, which until shortly before had been a vast region of largely unexplored wilderness. Beginning in 1881, the French attempted to build a canal connecting the Atlantic and Pacific through Panama, but the dangerous, difficult work and insurmountable financial problems caused them to abandon the effort by 1889. Fourteen years later, the United States cooperated with Panamanian revolutionaries, which led to a successful coup that separated the isthmus from Colombia and created the nation of Panama. These newly established diplomatic relations then made it possible for the U.S. to pick up where the French left off. Beginning in 1904, rivers were dammed, lakes created, enormous quantities of rocks and earth were moved, and locks were built. Once the canal was completed in 1914, surrounding forest was protected from cutting to ensure that water (and not rocks and earth) flowed unabated, keeping the waterway filled. Similarly, islands within the canal were protected, including the largest, Barro Colorado, which became a premier site for the study of tropical biology.

Indeed, the dean of American ornithology, Frank M. Chapman (inventor of the Christmas Bird Count, as an alternative to Christ-

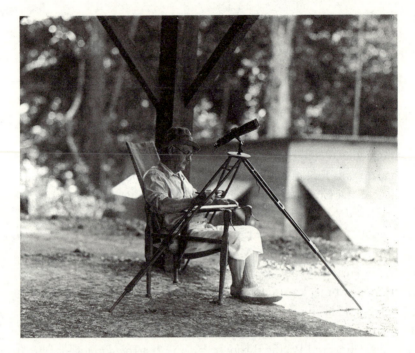

Fig. 1.1. Frank Chapman on Barro Colorado Island, circa 1920s. Image no. 277469. Courtesy of the American Museum of Natural History Library.

mas bird hunting), spent years on Barro Colorado (fig. 1.1). While living there, Chapman devoted a good bit of his time to studying the source of those raucous, snapping sounds—an amazing little forest bird then called Gould's manakin, and now generally known as the golden-collared manakin (*Manacus vitellinus vitellinus*).

This book is the story of that remarkable bird and all its wondrous biology.

The train still runs between Colón and Panama City, and wild sounds of the forest are muted, if in competition with engines for the dwindling acoustic space. Yet as each train recedes into the distance, the sounds of nature return, including the remarkable percussive snaps and "cheepoo" vocalizations of male manakins.

Those Exuberant Male Manakins

The wet season, having drenched central Panama since October, lifts around the beginning of January, and it is at this time that strange noises emanate from the lush, dense rainforest. These sounds resemble the snapping of human fingers, but are sharper, more explosive, like the cracks of many small whips or of firecrackers in a miniature Chinatown on New Year's Eve. Another peculiar sound is heard as well, much like the sound you would hear when running your finger along the teeth of a stiff plastic hair-comb next to a microphone with its amplifier and speaker turned up full blast. This cacophony is not heard everywhere, but emanates from discrete patches of forest, an abrupt and raucously chaotic symphony of percussion and song.

This is the sound of a group of male golden-collared manakins (previously called Gould's manakins) who have gathered to attract female manakins and then dance for and with them, all in hopes of gaining a copulation. Golden-collared manakins are quite common in many forests of Panama, ranging from seasonally dry forests in the south, near the Pacific Ocean, to quite wet forests in the north,

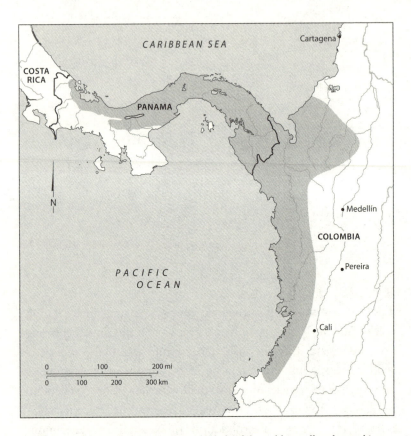

Fig. 2.1. Geographic distribution (shaded) of the golden-collared manakin (*Manacus vitellinus*). Map by Bill Nelson.

near the Caribbean Sea (fig. 2.1). To the west, closer to Costa Rica but also on the Pacific slope, a relative, the orange-collared mana-kin (*M. aurantiacus*), snaps and displays to orange-collared females. In the wetter northwest, another species called the white-collared manakin (*M. candei*) performs an almost identical display from western Panama and Costa Rica to central Mexico. Similarly noisy displays are performed throughout much of South America, from Colombia, Venezuela, and Trinidad in the north to Argentina and

Bolivia in the south, by a third related species, the white-bearded manakin (*M. manacus*).

Although these *Manacus* species differ in their plumage and in aspects of their ecology and behavior, they have many things in common, including elongated feathers that grow beneath their lower beaks. These feathers can be extended and retracted, much like the dewlap of many species of lizards. To early observers, these feathers resembled the beard growing from a human chin—thus, the birds are also known collectively as the "bearded" manakins.

Adult male golden-collared manakins are striking (see plates 1 and 2). Although only about 13 centimeters long from the tip of the beak to the end of the stubby tail, and weighing only about 18 grams, they have an outsized presence in the forest. Males sport a stark cap of black that sits atop the golden-yellow plumage of face, throat, and neck. These yellow neck feathers fully encircle the neck, forming a collar, giving the bird its common English name. The border between the black cap and the golden-yellow face passes strikingly through the bird's cold ebony eye and continues to the base of the bird's shiny blackish beak. The golden-yellow feathers of the neck merge at the upper chest into dull-olive green plumage that covers the whole belly. The feathers of the wings are a striking black, like the cap, with some black feathering extending across the upper back, separating the yellow collar from more olive-green feathering of the middle back. These drab green feathers extend down to the top of the clean, black tail feathers. Emerging from the olive-green belly are two strikingly red legs, thus completing the wardrobe of this attractive bird.

Adult females differ profoundly from their male counterparts in that they are fully covered in the dull olive-green plumage found on the belly and middle back of males. Their red legs, however, match those of their male partners. Juvenile males also possess this olive-green plumage (see plate 3), therefore looking exactly like females. Only during their third year of life, their second possible breeding season, do males acquire their adult plumage. The reason juvenile males resemble females is a matter of debate (to be discussed more fully in a later section), but it certainly makes it difficult to know with certainty when seeing a green bird in nature whether it is a female or a young male.

Returning to the sounds of the forest, we notice that, interspersed among the manakin snaps, we also hear various vocalizations, or calls, given largely by the males. The most common call is a relatively high-pitched "cheepoo," with the accent on the "poo." There are other similar calls as well, produced by both males and females. When a bird is upset or anxious, it produces an especially high-pitched and somewhat mournful "chee" call; when angry, it seems to say "peeyuk." Finally, as the males fly about, their wings produce a pronounced whirring sound. Thus, whenever a male flies within his lek, you know that he is moving, and you know where he is. Certainly, the other males are aware of this movement, and perhaps the females as well. Acoustically, then, on top of all the normal forest sounds of diverse birds and insects, the continuous dripping from the rain of the previous night, the breeze-driven rustling of leaves and palm fronds, the air around an active lek possesses a spectacular chorus of snaps and rollsnaps, cheepoos and chees, peeyuks and whirrs. There is a great deal of life in a golden-collared manakin lek (see plate 4).

What exactly is a lek, beyond a lot of noisy males moving about? Leks are aggregations of males that are actively courting females in hopes of obtaining a copulation or two. The first studied aggregations of courting males were of a kind of sandpiper, the Eurasian ruff. In this species, the breeding birds, like manakins, are strikingly sexually dimorphic, the males ornamented with beautiful, elongated black, brown, and white plumages, whereas the females (called reeves) look like any other relatively plain brownish-gray sandpiper. These birds breed in wet subarctic meadows. Males gather in groups, each with a small territory defended somewhat aggressively from other males. When an assemblage of dancing males was described by Swedish ethologists as "leks" (after the Swedish *lekställe*, or mating place), the name took hold. We now know that male animals of all sorts and all over the world gather together in leks, including a great diversity of bird species. Several types of manakin are among these, including our own golden-collared.

In addition to all this noise from the lek, the males' dances or routines are nothing less than spectacular athletic performances. The performance goes something like this: First, each male establishes one or a few "courts" in the forest. A court is the space between at least three small upright saplings approximately one centimeter in diameter and usually only one to a few meters tall. Between the saplings, the male clears, and keeps clear, a patch of the forest floor as bare dirt. This bare patch varies in size from that of a dinner plate to that of a decently proportioned serving platter (see plate 5). The three saplings around the cleared patch form a triangle, with one at each end of the clearing, and the third midway along one side. Although a male may create a few of these arenas, he usually prefers one, around which he does most of his dancing.

The dance itself, as illustrated in figure 2.2, involves the male jumping, not flying, from sapling to sapling, approximately one-half meter above the ground, and occasionally closer to the ground. The jump is spontaneous and fast, making the male appear like a projectile moving through the air. He holds his wings against his body, his beak is pointed forward, and his beard juts out, also in a forward direction. At the last millisecond or so before he reaches the intended sapling, the male makes a quick adjustment with his wings and extends his legs, opens his feet, and executes a perfect landing on the twig. In midair, the male rapidly and powerfully throws his wings over his back, and when the wings collide they produce the loud *snap* that we hear emanating from the forest. After a brief pause, the male jumps again to a different sapling, repeating the wingsnap and the impeccable landing. Sometimes, rather than jumping to a

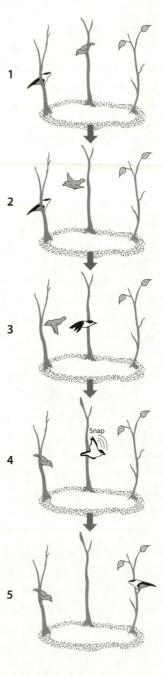

Fig. 2.2. A schematic (top to bottom) of the courtship dance of the male golden-collared manakin, in this case joined by a female, depicted here as the solid gray bird. Drawing by Bill Nelson.

sapling, the male jumps to the ground, doing a partial flip with a half-twist while also making a wingsnap. Upon hitting the ground, he springs up vertically using a "helicoptering" flight that produces a sound almost like the quack of an odd duck: the "grunt-jump" display. Overall, the male jumps about nine or ten times between saplings before stopping. Depending on various factors, like the presence or absence of a female, he may at this point depart the court, initiate another dance, or fly to a nearby perch for a rest. Or he might decide to perform a "rollsnap."

A rollsnap is the same basic maneuver as the single wingsnap, but rather than throwing his wings over his back such that his wings collide once while in midair, the male now throws his wings over his back repeatedly, some 10 to 23 times and at a speed of well over 50 beats per second, and he does this while perched. The sound this produces is as described above, like a finger run over a plastic comb, but very loud. Male manakins weigh only about 18 grams, about as much as three U.S. quarter-dollars. That so small a bird can produce wingsnaps and rollsnaps of such high amplitude is nothing less than astonishing.

A key feature of the display is its speed. The back-and-forth jumping of some males is so fast that, with the naked human eye, it is hard to tell just what is going on. You see a frenetic set of movements punctuated by loud snaps and calls, but the details of the display are imperceptible. Yet as we shall see, there is perfection in the movements that rival—or exceed—the greatest human gymnastic and dance performances.

Although males perform their dances and make their noises singly (or when a female is nearby), they are not alone. As mentioned previously, the birds have gathered together within a lek, a communal display ground where multiple males gather near one another; yet each retains a small "territory" defended from other males, within which he displays. The typical analogy used to describe a lek is the singles bar. Here, males gather seeking to attract a female. Although they assemble en masse, they may try to protect an open stool next to them at the bar, hoping a female will take a seat. A single male in a crowded bar is more likely to attract a female than he would in a lonely, otherwise empty bar. In other words, it is far better to take your chances in a group than going it alone. Moreover, many males in a bar can make a lot of noise—or might provide enough income for the bar to hire a band—and the noise would spill out into the street, attracting even more females. All increasing the likelihood that any one male might attract a female and, perhaps, start a romance.

Male manakins appear to do the same thing, except for a manakin the romance lasts a few seconds at most. Generally, anywhere from two to twenty males occupy an area of forest, which may average from 10 meters in diameter to half a hectare (about one acre) in size; rarely, they are even larger. The more males that gather, the louder the sound of snaps, rollsnaps, and cheepoos that emerge from the forest. Each male stays close to his own court or set of courts, and he is clearly aware of the other males around him and what those other males are doing. Upon seeing a female arrive at the lek, a male will rollsnap, usually several times. He will also descend to his favorite court and begin to display. Upon hearing this increase in activity, other nearby males will also perform rollsnaps and descend to their own courts to begin dancing. The increased energy in the whole lek is palpable as it explodes in sound.

Females appear to be picky, and will troop around the lek, observing and listening to the displays of a number of males until they identify one that pleases. The female then joins the male in his dance, but rather than jumping from perch to perch, she flies, somewhat awkwardly. If the female joins the male, there is a good chance she will let him copulate, which he does by doing a backflip with a half-twist to the ground, helicoptering up to the sapling above where the female is perched, and sliding down the sapling onto the female's back for a very brief cloacal kiss and the apparently instantaneous passage of sperm. With a ruffle of her feathers, the female departs, and the male returns to a position near his court where he can keep track of the activity of the nearby males in anticipation of the arrival of another potential mate.

Male manakin leks are open for business for extended periods of time. Adult males arrive at their courts and begin displaying almost daily in mid-January, continuing on through the dry season of February, March, and April and into the beginning of the rainy season in May and June. A few adults are still present from July to October, but the Panamanian forest gets really wet from October through December, and males can become noticeably scarce during these months. Nevertheless, except for flying off to feed on nearby trees or shrubs that bear small fruits or berries, for most of the year the males stay near their courtship arenas.

The daily manakin routine is fairly predictable. Shortly after dawn, when the forest light is sufficient for females to see the vivid yellow and black plumage of the males, the noisy work of displaying begins. After about two to three hours, silence falls. The males either go off to feed or will find a somewhat hidden spot in which to perch and rest during the hottest part of the day. Around two to three in the afternoon the males are at it again, and they continue to display until the light in the forest becomes subdued, at which point the displaying stops, their raucousness replaced by the calls of frogs and insects of all sorts. Although on some mornings or afternoons the males may seem absent, perhaps because a forest falcon or Jaguarundi cat is lurking nearby, as a rule the little patch of forest erupts like firecrackers twice each day for half the year or more.

These displays are a true natural wonder, impressive to just about everyone who experiences them. But these little birds are especially fascinating to biologists, arousing all sorts of questions. Why such a crazy display? How do the males make such loud sounds, and move so fast? Are there special features of their brains, spinal cords, muscles, or bones that allow them to do what they do? Do males learn how to perform the display, or does it just come naturally? Do they ever get tired of performing? Do female manakins see and hear the display as we do? Do they extract information from courtship displays that identifies the best males with whom to mate? How do males keep their courts clear, in the middle of fairly dense rainforest with big leaves from big trees falling all around? Such a commotion attracts female manakins and humans; does it also attract bird predators? Why are leks located where they are? Who chooses the site? Why do juvenile males resemble females? Why are males of Costa Rica and South America covered in many white feathers, whereas those of Panama are golden, orange, or even lemon colored? Can we learn something about ourselves from the study of these extraordinary birds?

These questions and more have inspired the work of many scientists. The studies by my lab over the past twenty-five years have provided some answers, and, together with the work of a cohort of biologists and ornithologists studying manakins and other species, both avian and nonavian, as well as collaborators with expertise in neurobiology, endocrinology, muscle physiology, bone morphology, bioacoustics, and aerodynamics, we are now beginning to piece together details of the complex creature that is a manakin. We will explore many of these facets of manakin biology in the following chapters.

Bear in mind that the various disciplines I just mentioned all interact to provide an understanding of the final organismal manakin product. The structures and processes at the heart of these disciplines both limit how manakins evolved and provide explanations for some of the remarkable attributes we observe. For example, the brain controls endocrine systems that produce hormones which impact the function of many tissues, including hormones that act back on the brain. In this way hormones can drive muscular systems by first modifying neuronal morphology and electrical properties that determine how hormone-dependent muscles contract. These muscles then signal back to the central nervous system to inform its response. Sounds are detected by minute structures in the ear, and the resulting signals are sent to the brain to drive cognitive processes, motor responses, and endocrine functioning. All of these inputs and outputs can influence gene expression patterns in many tissues that, in turn, further modify responses to inputs. Although this seems (and actually is) profoundly complex, it also, in a group of manakins, all assembles magnificently into a beautiful final product, nature's own *Swan Lake* or *Firebird*, in miniature.

THREE

Some Ornithology Basics

The biological wonder of golden-collared manakins comes into full view when we consider their various similarities to and differences from other birds. Thus, I will devote this chapter to a crash course in ornithology.

Manakins are an extraordinary group of small birds restricted to neotropical forests extending from Mexico to Argentina. The currently 53 named species of manakin belong to a single family, the Pipridae. This family resides within the very large order Passeriformes, or perching birds, a complex group some 6,000 species strong—60 percent of all the bird types on the planet. Virtually all of the small birds you see flitting about in trees or strolling on your backyard lawn are passerine birds. What they are not are doves or pigeons, gulls or sandpipers, hawks or eagles or falcons or owls, nor are they woodpeckers, cuckoos, hummingbirds, or swifts.

Most passerine birds belong to one of two suborders, the oscine songbirds (Passeri) and the suboscine birds (Tyranni). The oscine songbirds are the sparrows and warblers and jays and chickadees (tits), the swallows and nuthatches, thrushes and tanagers, starlings and blackbirds, the finches and wrens and mockingbirds. Even the crows and ravens are songbirds—big ones. Although there are about 2,000 species in the suboscine suborder, many of these birds are less familiar to us. They are the tyrant flycatchers, woodcreepers, cotingas, ovenbirds, antbirds, pittas—and manakins. Incidentally, many of the suboscine species live in South and Central America.

This is just a short list of all the Passeriforme birds, but I think you get the picture: there are many different kinds and they bear some great similarities in being relatively small and in being birds in or above forests and open ground. They are not, for example, birds of the open ocean. To a large extent they feed on insects, or on small fruits or seeds. Some are dull-colored and hard to see, like many sparrows and thrushes, whereas others bear brightly colored plumages and are quite conspicuous, like cardinals and many tanagers. Our awareness of the oscine/suboscine classification system will become evident when we focus more closely on what makes these birds different.

Of course, a vast number of characteristics make birds special, described in an enormous scientific and lay literature. To put it bluntly: birds are extraordinary organisms. I will not try to cover all of their biology in this brief space. What I will do is focus on those avian characteristics most applicable to golden-collared manakins and their anatomy, physiology, and behavior, features that make them worthy of this book. This includes a discussion of bird feathers, wings, and legs, some of their glands, and the brain regions involved in their ability to call and sing.

Although we now know that some, if not many, of the reptilian ancestors of birds were covered in true feathers or modified scales that resembled feathers, to most of us there is nothing more quintessentially "bird" than a feather. It is one of the most perfect structures in nature, incredibly light and pliant, coming in numerous shapes and sizes that accomplish a vast number of functions, impeccably. Feathers cover virtually the entire surface of the bird, varying in structure and size depending on location. Feathers provide insulation from bitter cold or extreme heat. They provide color and pattern, whether for camouflage or to attract attention, by packaging pigment or by being hollow or by possessing crystal-like nanostructures that create iridescence. Feathers can be impenetrably waterproof; they can be noisy or perfectly silent; when ruffled, they can be repaired by zipping their interlocking barbs and barbules; they can function as flaps, rudders, and airfoils that slice through the earth's atmosphere like little else on the planet.

Bearing all this in mind, you can see that just about any discussion of a bird's biology requires some attention to feathers. This is certainly the case for manakins, as we will amply see throughout this book. Because really, what's better than a feather?

Of course, what distinguishes birds from most other vertebrate animals is their ability to fly, often with extraordinary skill and grace. Flight comes in many forms, but the basic elements involve lift—that is, vertical movement against gravity—and horizontal movement, only rarely not in a forward direction. Lift can occur when air is deflected upward (through convection) by landforms such as mountains, or even by buildings, or when rising columns of heated air push the bird, its wings outspread, aloft. This strategy is used by many large birds such as hawks, vultures, storks, and pelicans, but also by many small birds such as swallows and swifts. Lift is also created by air moving across the outstretched wing's surface, a phenomenon described as Bernoulli's Principle. The curvature of the bird's wings and flight feathers is such that the air moves a greater distance over the top of the wing than under it; because all this air is moving together at the same time, the density of the air is greater beneath the wing than above, and it is this difference in pressure that pushes the wing (and the bird) upward. This lift can be created when a bird simply opens its wings against the wind (think of a cliff-dwelling albatross taking flight) or when the bird flaps its wings.

Flapping flight is quite complicated, for the bird uses its wings not only to push air downward, thus gaining lift, but also to create airflow by pulling the wings forward, producing a brief moment of differential pressure and additional lift. Many birds are exceptionally light, with very low body mass, and for them the lift required to move against gravity and gain forward momentum is much less than for a heavier bird. Even so, flight requires strength, whether it's for a short burst—think of a heavy wild turkey—or it's to cover extraordinarily long distances for extended periods of time, as with the vast number of migratory birds that traverse thousands of kilometers in a single exhausting flight.

Feathers work with the structural components of a bird's body to support the bird's weight against the upward force of lift and downward force of gravity. They also work to reduce drag, which inhibits lift. Not surprisingly, the bird's body itself can create drag, so birds are aerodynamically shaped, the body feathers providing a sleek, smooth face against oncoming airflow.

Many features converge to make most birds flightworthy. Imagine a bird in flight with the beak-to-tail axis horizontal, the bird's chest and belly (the ventral surface) facing downward and the bird's back (the dorsal surface) facing upward; wing strokes then occur in an up-and-down motion, with up being extension and down being retraction.

Flying birds function best without excess weight, and what weight they must carry needs to be positioned so as to enable lift and balance. That means carrying most of their weight beneath their wings: that is, when airborne, their center of gravity is best positioned centrally beneath the load-bearing wings. To achieve this, birds have evolved several key features, listed here in no particular order. Rather than teeth, birds possess a lightweight beak situated far from the center of gravity. To "chew" their food, they possess a muscular gizzard positioned at the top of their abdomen. Birds need powerful flight muscles to lift and retract their wings. The pectoralis muscles that depress the wings (pull them down) are located ventrally, on their chests, much like our own "pec" muscles. We can lift our arms behind or above our backs by contracting muscles *on* our backs. Unlike us, birds lift their wings using another pair of ventrally located muscles. When contracting, these muscles (the supracoracoideus) are able to "lift" the wings because they connect to the dorsal side of the bird's humerus via a tendon that loops over a pulley system located in the shoulder. Hence, both of the large and powerful wing extensor and depressor muscles are located beneath the wings of our horizontal bird described above, reinforcing the center of gravity.

Another anatomical specialization to reduce weight is via the bird's skeletal system. Solid bones are heavy, but birds are able to retain incredible skeletal strength by means of modified, hollow bones. Some bones of some species are even connected to the respiratory system, allowing inspired air to make its way into and reside within the skeleton for a period of time before being exhaled. This one-way flow of air through the lungs is an evolutionary legacy of reptiles that increases oxygen uptake and CO_2 release. While not directly involved in weight reduction, it is still a cool physiological trick.

As we will see in later chapters, many of the features that make a bird a bird, among them flight muscles, flight feathers, and wing bones, have been subject to modification in golden-collared manakins, enabling them to perform their elaborate courtship displays and attract females for mating.

Bipedalism is something we take for granted in birds, given their remarkable ability of flight. Of course, we see ourselves as the masters of bipedalism: our need for specialized hands to make tools and manipulate our environment meant that we had to quit swinging around in trees and stand upright. But the birds evolved into bipedalism many millions of years before us, because they needed specialized arms and hands to provide lift and propulsion—that is, to fly.

Large numbers of birds walk, hop, or run with exceptional grace, speed, and agility. In most cases, these birds also fly, and usually quite well. Conversely, there are many species in which flight is so important that legs have become somewhat secondary, such as hummingbirds and swifts, which have tiny legs that can do little more than perch, and sometimes just barely. The legs of some species, such as loons and grebes and some ducks, are so highly evolved for swimming that walking is for the most part forsaken. These birds can indeed patter across the water to take off for flight, but such hindleg-propelled locomotion would be impossible for them to perform on land. Of course, some birds are flightless—penguins, for example. But some, such as ostriches, rheas, emus, and cassowaries, are superbly adapted for terrestrial locomotion; with hindlimbs and muscular and skeletal adaptations that should be the envy of the bipedal world, these amazing runners receive less attention than they deserve for their prowess.

Then there are the legs of some of the best flying birds. Eagles, hawks, falcons, owls, and other birds of prey have hindlimbs modified specifically for capturing and carrying prey. I urge anyone to examine the legs of a female harpy eagle, the largest bird of prey in the world, and not say, "Holy Toledo!" Each leg, as thick as a medium-sized man's arm, powers five long and thick razor-sharp talons. These are hindlimbs to fear. The legs of most birds, in fact, are especially good at grasping, not necessarily to crush and kill their quarry, but simply to perch. Just look outside on a windy day and you may find a bird comfortably perched in a tree, paying no mind at all as the surrounding branches are whipped around left and right, up and down. Even relatively large and heavy birds that spend a lot of time walking around, like pigeons, can perch comfortably on a tree branch under such conditions.

On top of these essential hindlimb functions, some birds use their legs to dance. It should be no surprise that male ostriches, with their especially strong legs, dance around to attract females for copulation or pair-bonding, but the same is true of many other species, including the African, Australian, and Eurasian bustards and many galliform species, such as turkeys, grouse, and pheasants. Cranes are especially graceful dancers. The list of dancing birds is long, and the agility to move about on the ground, to jump, to run, to twist and turn, to squat and stand, requires unique modifications of the hindlimbs for strength, dexterity, and stamina. But it's all worth it to entice females.

Like these other birds, manakins are dancers! Golden-collared manakin males "dance" around their arenas, largely by jumping from sapling to sapling or from a sapling to the ground and back, often adding a flip with a half-twist when jumping to the ground. The jumps, from a vertical branch, can be up to a meter long. Relative to body size, that would be like you jumping 15 meters or more, without a running start—never mind hanging sideways on a tree to start with. When perched, the males often lean forward somewhat awkwardly, showing off their forward-facing beard of feathers to nearby females. Sure, the birds are fairly light, but still, this stance is impressive. It is difficult to find a human analogy, since we cannot grasp with our feet, but think of a male gymnast performing an iron cross on the rings, a wholly unnatural position that requires exceptional strength, as seen in his bulging lats, pecs, and deltoids. Although the fine covering of feathers renders male manakin leg strength invisible to us voyeurs of manakin behavior, manakin leg muscles are equally impressive and worthy of our respect—as we will see in a later section.

So great legs allow hindlimb-based locomotion or hunting or perching or, of greatest interest for us, dancing. But adaptations that promote greater leg function and use also liberate the forelimbs to evolve to do whatever else the organism needs. In birds, they have evolved into the magical structures we call wings, giving birds the remarkable accomplishment of flight and enabling them to claim the air like no other animal that has ever occupied planet Earth. But wings do more than just enable flight.

It seems almost heretical to say that this book about birds has almost nothing to do with flight. But it is true, and we will spend little time focused on this trait. Manakins do indeed fly, and quite well and quite fast, though never very far. Indeed, an average-sized river is sufficiently wide to limit whole species of manakins from crossing. Where there is no gene flow, species can evolve, and this is what has happened in western Panama, where golden-collared and white-collared manakins (both in the *Manacus* genus) are physically separated. What is especially important to note here is that although manakins may have forsaken evolution of some flight capabilities, they have evolved wings for uses above and beyond flight, namely for courtship.

Many other species of birds have also traded wing adaptations for flight for alternate uses. In penguins, cormorants, and distantly related seabirds such as shearwaters, alcids, the boobies, and gannets, and even in a songbird, the dipper, the forearms evolved for efficient swimming as well. Some recent evidence suggests that even starlings can swim if forced to, and do so using their wings. Of course, penguins' compact forearm flippers represent an extreme form of specialization, and one cormorant also gave up flight completely upon adapting their forearms for swimming. Yet many of the species described above are superb fliers, in addition to using their wings to swim.

But wings do more than allow locomotion.

Many years ago my family visited Kalamazoo, Michigan, to attend my brother's PhD commencement from Western Michigan University. During our stay, we paid a visit to the Kellogg Bird Sanctuary. At that time, and maybe still to this day, the sanctuary had compounds with exotic bird species that could be approached relatively closely. I went into my bird know-it-all mode and began talking about the different species we saw. One exhibit held several black swans (*Cygnus atratus*), native to Australia. Whether white or black, swans are beautiful, impressive birds. But they are also large and, frankly, nasty. Many children who have tried to feed a peaceful-looking mute swan (*C. olor*), introduced from Eurasia and now occupying many parks across North America, have undergone a frightening experience, as swans are very prickly and known to attack.

While I stood in front of the black swan enclosure lecturing on some aspect of swan biology, my hand crossed through the fence, just a few centimeters from the head of a swan. In an instant, the swan swung its wing powerfully upward, striking my finger with its wrist, and then retracted its wing back to its side, as if nothing had happened. The strike was so fast, most people didn't see it. *I* barely saw it. But I sure felt it. I thought it had broken my finger, the pain was so intense. The force that had hit me was remarkable, not only for its power, but for the incredible speed and accuracy. Had I been a fox or coyote, or a dingo in Australia, I would absolutely think twice before approaching that swan, or its nest, or any swan or its nest, ever again.

As it was, I kept on talking as if nothing had happened, too embarrassed to admit that I'd been stung by the very animal about which I was pontificating. My finger swelled and hurt for days.

The point of this story, for our purposes here, is that bipedalism liberates the wings to do many things. Although flight is a primary use of wings, for the swan, wings are also weapons. The black egret of Africa folds its wings up and over its head as it wades patiently in shallow water. The shade thus created attracts small fish, which in turn are devoured by the bird. Many birds, including the familiar northern mockingbird, do something similar, rapidly flashing their wings as they hop in short grass to scare insect prey, which the bird snaps up as the bugs try to escape. An unusual bird of the American tropics called a twistwing performs a variation on this, lifting first one wing over its back, then the other, presumably also to startle insect prey. Pigeons and many doves are well known for clapping their wings over their backs upon takeoff, both to signal other pigeons in the flock to take flight and likely to alert them to potential danger. The wings of many birds, especially some ducks like goldeneyes, emit whistling sounds, probably also to communicate positional information to other goldeneyes in flight. These are but a few examples of wings employed in ways outside of their primary function for flight.

Perhaps the most conspicuous nonflight function of birds' wings is in courtship rituals, which can involve both visual and auditory signaling. In this, manakins are a special case, for their wings are used to make extraordinary sounds when they are courting females and defending their arenas.

Many manakins employ an extraordinary array of audible signals, both vocalizations and sounds created by movements of their wings or tails. To fully appreciate manakin sound production, it is useful to consider the numerous specializations that have evolved in birds more generally.

Most people consider bird sounds to be an ever-present feature of the natural world. Many birds make noises, whether by their wings and tails, by specialized sacs, by rattling their beaks or hammering on trees, but the most familiar bird sounds are their calls and songs. These are made by the movement of air across membranes located in the trachea, which are stretched and relaxed by a small set of exquisitely controlled and highly specialized muscles. This vocal structure, called the syrinx, is somewhat analogous to our own larynx. By regulating airflow across syringeal membranes (by regulation of the respiratory system), by impeccable control of these syringeal membranes (by neural control of hindbrain motor neurons that inner-vate these muscles), by precise maneuvering of the neck and beak, birds have become masterful vocalists.

Most birds vocalize, and we tend to divide these vocalizations into two broad categories, simple calls and complex songs, though there is extensive overlap in their form and function. Both can attract mates or repel rivals, and convey perceptions of the environment—for example, alerting others to danger or to a source of food. Exactly what constitutes a call versus a song is a subject of considerable debate, but for our purposes we will stick to a relatively straightfor-ward definition, with calls being given by almost all birds, including manakins, and songs being produced by that group of birds known as the oscine songbirds.

Some of the most beautiful sounds in nature are sung by this group of birds. If you have never sat quietly in a North American forest and heard the echoing song of the winter wren or the melan-choly songs of the hermit or Swainson's or wood thrushes, you are missing some of the most mesmerizing auditory experiences in na-ture. If you are lucky enough to awaken to the dawn chorus of a New Guinea forest filled with songbirds, you will be rewarded with otherworldly magic.

A number of anatomical traits separate the oscine and suboscine birds. For our purposes, the important differences lie in the struc-ture of their syrinx, the song-producing organ; in the properties of their inner ear, which hears songs; and in the brain regions that pro-cess incoming song information or that can learn song and control the muscles of the syrinx and respiratory system to produce song.

Within the huge community of neuroscientists across the planet (some 30,000-plus attend the annual Society for Neuroscience meet-ing), the brain of the oscine songbird is arguably one of the best understood complex neurobiological systems of all. You might won-der why. There are many reasons, but a few stand out.

First, like humans, oscine songbirds must learn their song. While they may possess a rudimentary ability, to produce a true song, with all its rich syllables, tones, amplitude, and patterning, the young birds must hear it, from their father or from a tutor, and store it in the brain. At a later age, the birds then begin to babble. They produce a diversity of sounds, and as they make sounds that match those memorized from their father (or tutor), those sounds become encoded in the bird's song routine. Eventually the babbling becomes subsong, and ultimately song itself—a perfect match of the father's.

Within species that breed in the northern hemisphere, the baby bird might hear the father's song in July, during the breeding season, but then might not begin babbling until the following springtime, in April, at which time the song can become fully crystallized. Thus, the memory for the song is stored, perfectly intact, for a full eight months.

It's hard to convey just what an incredible accomplishment that is. But let me try to place this in a human perspective. Imagine an American child born in a foreign hospital, where the nurses and the ob-gyn all speak a language different from that of the mom and dad. Let's say, French. The newborn hears, "Ahhh! Un joli garçon . . . le premier souffle était bon et fort . . . nettoyons le bébé et rendons-le à sa mère" ("Ahhh! A beautiful boy . . . his first breath was good and strong . . . let's clean up the baby and give him back to his mother"). The family goes back home, the boy grows up, and as a young man he suddenly, seemingly out of nowhere, starts emitting strange, French-sounding phrases. His parents are worried. After a few weeks of this, the young man suddenly says, with a perfect Loire Valley accent, "Ahhh! Un joli garçon . . . le premier souffle était bon et fort . . . nettoyons le bébé et rendons-le à sa mère." The boy had encoded an acoustic memory and held it intact for a long period of time (in the human case, years) before suddenly speaking (vocalizing) sounds similar to the long-held acoustic memory, the result being a perfect copy of the words of the French ob-gyn.

In this example, the newborn is hearing words, akin to the syllables of the bird's song. As the young adult, he says those words *exactly* as the ob-gyn said them, as if the voice of the ob-gyn had been recorded on a high-quality reel-to-reel and was now played back on high-quality speakers—but this perfect match is coming not from a sound system but from the mouth of the now grown-up baby. He had to neurally encode (memorize) the words and all of the subtle acoustic properties of each of the sounds perfectly (like the tape recorder), hold that memory intact for years, and then re-create those sounds from his own larynx and mouth (like the speakers) as a perfect match to the ob-gyn. Some humans actually do perform such monumental tasks of memory and playback, and we call them savants. In that sense, every male (and in some cases female) songbird is a savant!

It is important to realize that song itself is the result of the co-ordinated contractions of the tiny muscles of the bird's syrinx that stretch or relax membranes of the bird's trachea, such that airflow across these membranes makes sound. The output of the motor neurons connected to those muscles must be controlled very precisely. In addition, the firing of those motor neurons must be coordinated with muscles involved in respiration, in the opening of the beak and stretching of the neck, and indeed, even in the posture of the bird on its perch (or while in flight). This motor control is guided by so-called premotor neurons. It is the firing of these neurons that orchestrates the many signals that induce the downstream motor neurons to fire in patterns that cause the muscles to contract in correct time and amount to produce all the elements that we hear as song.

A large number of brain regions contribute to the songbird's ability to learn and memorize song and to then actually sing. A few regions stand out for their importance (fig. 3.1). These include clusters of neurons, or nuclei, called HVC and RA, regions that drive the motor output of song, as well as lMAN, Area X, and DLM, regions

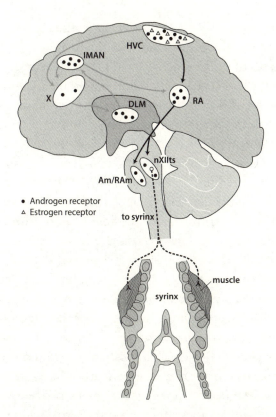

Fig. 3.1. A cross-section of the brain of a songbird (facing left) that illustrates a few of the more significant regions involved in the learning and production of song, including neural connections, expression of sex-steroid receptors, and projections to the muscles of the bird's syrinx. Drawing by Bill Nelson.

where songs are learned, memorized, and continuously honed. (Note that several of the brain regions once had more lengthy names but their acronyms are now in common usage.) Most of these regions are interconnected by neural projections. For example, HVC projects to RA, which also receives projections from lMAN. RA projects to two regions in the bird's hindbrain, AM/RAm, involved in respiration, and a nucleus called nXIIts, which in turn sends projections to the muscles of the syrinx, the bird's vocal organ.

The songbird syrinx is a collection of muscles surrounding the site where the bird's trachea splits into two bronchi. Contraction of these muscles, as driven by neural signals from motoneurons located in nXIIts, which are fully coordinated in timing and intensity by HVC and RA, produce song as air passes through membranes lining the syrinx. This entire neural system is a wonderful example of what is required for a vertebrate animal to perform a complex behavior. Given that birdsong is often used by males to attract females and claim territories for breeding, it is not surprising that some of these brain regions are sensitive to sex hormones. When considering all of this neurobiology, recognize also that a vast number of other neural systems come into play. Birds must hear song, so auditory inputs are essential. Moreover, the bird must be motivated to sing, so neural centers involved in sexual or aggressive drive communicate with song motor systems. All of these clusters of neurons are exquisitely coordinated in their firing patterns to achieve the remarkable behavior that we experience as birdsong.

This whole process represents a phenomenal neurobiological capability, relying to a great extent on the bird's auditory system, including an ability to identify and isolate song from lots of other sounds in nature. As in all vertebrates, songs cause hair cells in the inner ear of the bird to vibrate, sending signals to the bird's midbrain; these in turn project to the thalamus and higher cortical centers that process the information, to tell friend from foe, conspecific from nonconspecific, male from female, and, in juveniles, whether it's dad's song and should be memorized for performance at a later time. Many male birds can identify the discrete features of the songs of their neighbors of the same species—their conspecific rivals—and, by assessing the amplitude of those songs, determine where in space the neighbor is located. If that neighbor is too close to the territorial boundary, the male responds with an aggressive confrontation, often involving lots of posturing and, of course, singing.

Finally, input from this auditory system must intersect with the motor output system and with memory storage sites in the brain. A special region of the brain receives processed auditory input and projects both to premotor circuits and to the song learning centers. This whole intricate sensorimotor learning system in oscine songbirds is called the song control system. Interestingly, whereas it is highly developed in the oscine songbirds, rudiments of the system have been found in non-songbirds like hummingbirds and parrots.

Why this detailed description of the song system of oscine songbirds? If the vast number of brain regions and all their connectivity is needed to vibrate a few tiny muscles in the birds' throats to create song, just what is going on in the brains of manakins to produce not only their vocalizations but also all their other forms of acoustic signaling, the roll-snaps, for example, that require contractions of numerous rather large muscle systems moving their wings? Now, remember that the males also communicate visually, performing their remarkably complex courtship dances that require the perfect timing and contraction states of a vast number of large and small skeletal muscles. Some very special neural circuits live within the manakin brain. Manakins must have evolved their own specialized neural circuits that underlie their acoustic and visual signaling. Yet these circuits likely receive projections from neuronal systems that exist in virtually all other birds, namely those that motivate sex behavior and that are crucial for some vocal behaviors.

Recall also that each species of manakin has its own unique physical courtship display. Thus, males of each species have evolved their own neuronal systems driving the movements of this or that wing or leg or neck muscle and doing so quickly or slowly and simultaneously or independently. At the same time, the birds are calling, which requires appropriate and coordinated firing of discrete neural circuits innervating syringeal and respiratory muscles. We are just beginning to understand some of these elaborate neuromuscular systems in the golden-collared manakin, and what we have found is indeed impressive.

FOUR

The Ways and Means of Wingsnapping

The wing- and rollsnappings of male manakins are extraordinary behaviors, not only because the snaps are so loud for such a small bird, but because they are produced by the wings. It is difficult to quantify the amplitude of such sounds unless conditions are just right for their measurement. But we know these snaps are loud because they can be heard at a distance in forest, despite the dense foliage through which the sound must pass. Frank Chapman, who first studied golden-collared manakins almost a century ago, said he could hear them at 300 meters across open water.

Acoustic analysis of the snaps shows them to be a broadband noise, with the principal energy at around 5000 Hz (5 kHz, or 5,000 cycles per second). This differs from, say, a pure tone or whistle, where virtually all the energy is devoted to a small frequency range. For example, although a steady human whistle on a single note might have its fundamental energy anywhere within the range 500 to 5000 Hz, it will be possible to measure a specific frequency for any given whistle—say, 3520 Hz, equivalent to the highest A note on the piano—which we perceive as a pure tone. In the case of a

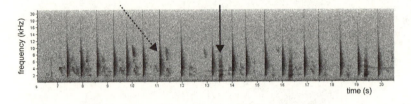

Fig. 4.1. A sound spectrogram of a display sequence of a male golden-collared manakin. Provided by Leo Fusani and reproduced with permission.

snap, although it may be possible to measure the peak energy at, say, 5000 Hz, the sound's full range of energy is spread uniformly across all the frequencies audible by the human ear; thus, rather than being perceived as a distinct note or tone, it comes across as broadband noise—a loud click or snap. You wouldn't say, "Hey, that snap was in the key of C," because the snap has no real "key" or "note." Adding to the effect is the fact that all that energy is momentary, lasting only a few milliseconds (or more if there are echoes), unlike a pure tone or whistle.

Scientists typically analyze sounds, or acoustic signals, by creating sonograms. Let's look at the sonogram of a manakin display sequence (fig. 4.1). This display lasts about 20 seconds, as shown on the bottom axis. The frequency spectrum of the sounds recorded is depicted on the left axis, ranging from 0 to 20 kHz. The darkness of each signal corresponds to the amplitude or intensity of the sounds. The dashed arrow points to a typical single wingsnap. The solid arrow depicts a "grunt" of the grunt-jump display element. In between many of the "snaps" are cheepoos produced by neighboring males in the lek. You can see that the snaps have a peak energy around 5 kHz, but at least some sound energy is produced continuously from below 1 kHz to, in a few cases, above 20 kHz. Note that cheepoos last a comparatively long time (almost half a second each), with their primary energy (called the fundamental frequency) sweeping downward from approximately 4 kHz to 2 kHz. Note also the energy signals at approximately 5 and 8 kHz (for the chee note) and 4, 7, and 9 kHz (for the poo note), which are harmonics.

Both single snaps and rollsnaps are similar in sound structure, but rollsnaps are remarkable in that the snaps are produced in such rapid succession. A typical complete rollsnap, consisting of about 15 individual snaps, lasts about 300 milliseconds (about one-third of a second). We recorded one male who produced 23 snaps in a row, over the course of almost a full half-second. Each individual snap is separated from the next by a bit over 20 milliseconds. And all of this is produced by the wings!

How can the wings produce such a perfect, loud sound, and do so singly, in midair, or repeatedly, at great speed, with a perched individual? Back in 1935, Chapman noticed that the wing feathers of the males were shaped differently from those of females and also from other manakin species, so he assumed it was the feathers striking one another that created the sound. Yet he was puzzled. "It is difficult to believe," he wrote, "that so loud and so hard a noise as the snap can be produced by an object as soft as a feather. Doubtless, for this reason, it has often been attributed to a snapping of the mandibles. But if this were true we should expect to find the bill of the male heavier or otherwise different from that of the non-snapping female. Furthermore, if the snap were made by the bill it would not, presumably, be necessary to jump when snapping. On the other hand, the somewhat large, slightly curved and stiffened shafts and broad webs of the secondaries in the male are doubtless of functional value in producing the snap."[1]

To begin to understand how the manakin snaps his wings, it is first important to consider just what a "snap" is. How does one make such a sound? Well, there are a surprising number of ways. You can, of course, "snap" your fingers by pushing the tip of your thumb onto the tip of your middle finger and letting go, allowing the thumb to release toward your index finger while the middle finger strikes the fleshy base of your thumb. One kind of "snapping" sound is easily created using a belt, preferably a relatively wide, strong leather belt. Hold the buckle and the belt's end in one hand, creating a complete loop. Grab the free rounded side of the belt in the other hand, bring your hands together, squishing the loop vertically, and then rapidly pull your hands apart to let the two sides of the belt "snap" together, producing a sound of the same name. I have demonstrated this in scientific talks about manakins, and you can imagine the surprise of the audience when, while on stage, I calmly whip my belt off of my pants. The hand clap we use to demonstrate appreciation, usually when we are in a crowd, is not such a different sound. You can take a small to medium-sized wooden stick and swiftly break it in two to create a sharp "snap" or "crack" sound. We can do many things to create loud snaps, but of course we are large animals. Let's say that the average human weighs 150 pounds, or just over 68,000 grams. The average male manakin weighs about 18 grams; so we are, on average, 3,780 times larger than a male manakin. Most of the sounds we make (without help from external devices) are not 3,780 times louder than a manakin's wingsnap. How do they do it?

Before we tackle that question, let's consider for a moment just what is happening when we make our snaps. Sound originates when something creates alternating pressure waves that move outward through a medium like air or water. The question then really becomes, what causes these waves? There are three accepted mechanisms. One involves the vibration of a body of some sort—the beating of a drum or the plucking of a guitar string, or, for a woodpecker, hammering against a tree trunk. The drumhead, the string, and the tree all vibrate upon impact, and that vibration alternately pushing against the atmosphere creates waves that move horizontally through the air until they hit your ear. In the ear, the pressure wave sets up a vibration of membranes that are transmitted to tiny hair cells (not actually made of hair, but tiny fibrils that emanate from the membranes of specialized cells and thus resembling hair) in the cochlea that in turn vibrate and tell the brain that sound has been received.

Sound can also be generated when air becomes highly unstable, such as when it is forced out of a jet engine at very high velocity. Sound is also produced when an object moves through the air faster than the speed of sound. Some of us are familiar with sonic booms, especially those of us who live near a military airbase or have made a visit to NASA's Kennedy Space Center or one of any number of now-private rocket launching facilities. These tremendous explosive (really, compressive) booms occur when an aircraft flies faster than Mach 1, the speed of sound (1225 km/hr or 761.2 mi/hr). Things smaller than aircraft can move through the atmosphere at such supersonic speeds as well, as when a bullet is fired from a gun. A whip "cracks" upon being flung outward and then rapidly retracted; the sound is produced as the wave of leather at the tip momentarily breaks the sound barrier.

Let's return to the examples of us humans producing snaps or cracks—the finger snap, the belt snap, the hand clap, and the breaking stick. The first three examples are all somewhat similar. The action of rapidly bringing two solid (or semisolid) objects together causes air to be rapidly displaced, or ejected outward. The ejected air is at high pressure, but leaves behind low pressure. It is this pressure wave that you hear as sound. The sound can be modified by changing the way the ejected air is created; for example, a clapping sound can be altered by changing the curvature of the palms. Although the tissues of the fingers or palm may also vibrate upon impact and produce some noise, the bulk of the energy that produces the clapping sound comes from the air being momentarily ejected by the act of applause. The last example, the breaking of a stick, produces a momentary loud "crack" because the stick itself vibrates when the explosive break occurs. Sticks of different sizes and robustness produce different sounds because they require different amounts of momentary energy (the energy needed to achieve breaking pressure) and, simply, because different dimensions and materials will have different responses.

So, back to the question of how a male manakin makes his wing-snaps. In fact, the behavior occurs so swiftly it is hard to tell just what's going on. Is the sound due to the vibration of the body, to air being rapidly expelled, or are the wings moving at supersonic speeds?

Working alongside Ted Goslow, an expert on muscle physiology and bird locomotion (and at the time a professor at Brown University, but now retired), and Leonida Fusani, a postdoctoral fellow in my lab (now a professor at the University of Vienna and director of the Konrad Lorenz Institute of Ethology at the University of Veterinary Medicine in Vienna), we tried to explore this question using ultra-high-speed videos of wild male manakins. Ted's equipment had the capability of obtaining a brief eight-second sequence of a manakin behaving. However, the camera and all its accessories were hardly designed for field work. First, the camera and computer were powered by car batteries, encased in a sturdy yet heavy crate, and the charge only lasted an hour or so before requiring an overnight recharging session. A separate case held the computer, and a third case held the digital audio and video recording devices and accessory equipment. We then had to schlep the three heavy crates into the forest, set up the apparatus on a court occupied by a known male, stand back about 10 meters, and hope the male arrived and was cooperative. To obtain valuable images that would document the details of the male's wingsnap or rollsnap, the camera had to be focused on a small region of the court, or on a perch where we had seen a male rollsnap previously. If a male did arrive, we had to hope he performed at precisely the spot we had predicted. If it looked like the male was indeed snapping at the appropriate spot, we would capture the image with a hand-held trigger. In the vast majority of cases, however, the male was slightly out of the frame. After a few shots, the batteries were drained, and our day was finished. All the equipment then had to be schlepped back out of the forest.

Adding to our difficulties, the equipment was not waterproof, and we were working in rainforest. When weather threatened, which it did almost daily, we had to quickly close up shop and, lugging our heavy crates, run out of the forest and back to a car sometimes several hundred meters away. Although the three of us were in fairly good shape, we were not spring chickens, so this whole set of experiments was exhausting and often frustrating. We did obtain some excellent sequences, though, some of which I still show when I give public talks about manakins. However, we gained no images that allowed for the kind of analysis we wanted.

Unknown to us at the time, advances in high-speed videography had produced more portable cameras and recording devices, and two other manakin investigators, Kim Bostwick (now at Cornell University) and Rick Prum (at Yale University), had taken advantage of this equipment to do some of what we hoped to achieve. Using high-speed video, they recorded wingsnapping in species very closely related to the golden-collared manakin: the white-collared manakin (*Manacus candei*), which lives in Central America, largely west of Panama, and the white-bearded manakin (*M. manacus*), with an enormous range throughout South America. They were also somewhat smarter than us in that they recorded some of their birds in the relatively dry, predictable environment of the San Diego Zoo and Wild Animal Park. These investigators examined not only individual wingsnaps and rollsnaps, but also several other kinds of sonations—that is, nonvocal sounds produced by means of wings, legs, and feathers.

For the single wingsnap (which, you will recall, is performed midair during the jump-snap display), Bostwick and Prum observed that the male "forcefully claps together the dorsal surfaces of the wrist joints, approximately where the bases of the remiges insert. . . . [T]he *snap* is produced using a wing-against-wing clap of the dorsal surfaces of the wings above the male's body." They concluded, "Contact between two solid structures is a minimum critical prediction for a percussive mechanism. Extremely rapid motion of a structure through air is necessary for either the whip or vacuum mechanisms, with the whip mechanism requiring motions faster than the speed of sound. The absence of these respective actions during sound production amounts to rejection of these respective hypotheses."[2] Their observations were very interesting and could not have been made without the use of high-speed photography. Yet they still could not explain just how such a small bird could produce such a loud sound, even by forcefully throwing its wings together.

By this time, we had wised up and purchased our own portable high-speed camera for an independent set of studies of the dance elements of the manakin courtship display. Leo Fusani had managed to precisely synchronize audio with the video input collected from independent recording devices, and we made further modifications to correct for the differential speeds of light and sound in our synchronization (we did this by making high-speed video-audio recordings of our friend Duane "Duano" Keith playing spoons against the palm of his hand in our apartment in the rainforest village of Gamboa, Panama). With this in hand, and together with a new PhD student, Juli Barske, from Germany, we set out to make score upon score of recordings of courting male manakins in the field, as well as some in captivity.

We also obtained numerous videos of males wingsnapping from different angles. One high-speed recording of a wingsnap is shown in figure 4.2. In this case it is a wild male manakin captured in mid-air using a camera recording at 1,000 frames per second. (Although the images appear somewhat blurry, bear in mind the difficulty of capturing such an event performed by a wild bird in a relatively dark forest at such a fast shutter speed.) Looking down on the bird (ignore the black square), one can see the bird's wings raised over his back; as they come together, what appears as a black triangle in frame 1 (top left) gets smaller in frame 2 (top right) when the wrists strike one another. The triangle then flattens out as the bird starts to pull his wings apart in frames 3 and 4 (bottom left and right, re-

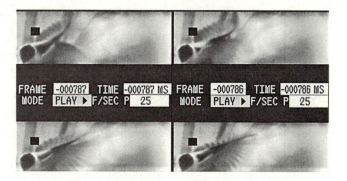

Fig. 4.2. High-speed video capture of a wild adult male manakin
(as seen from above) performing a wingsnap (top right image)
while in midair. Photos by author.

spectively). The bird's primary and secondary wing feathers are a
blur in the background of each image, though you can see them
splayed out above and below each wing in each frame. This observation confirmed previous work of Bostwick and Prum showing that,
indeed, the wings do collide at the top of the wing stroke. One can
clearly see here that it is the wrists that impact one another.

You can picture this a bit differently by trying to perform a wingsnap yourself. Hold your arms straight in front of you and try to
hold your wrists together. Imagine that your fingers are feathers
and bend your wrists so that your fingers point away from one another. Now imagine what it would be like to powerfully hit your
wrists together. As a next step in this exercise, try to assume this
same position with your arms held straight behind your back. You
actually cannot make your wrists contact one another, unless you
are a lot more flexible than I am. But that is what the male manakin
does: without being able to see what he is doing, the male throws
his wrists together perfectly each time, even when doing so impeccably during a rollsnap.

While in Panama collecting these video images, we ran into

Martin Wikelski, now an honorary professor at the University of Konstanz in Germany and the director of the Department of Migration of the Max Planck Institute of Animal Behavior (previously a part of the Max Plank Institute for Ornithology), who introduced us to George Swenson. Martin had done postdoctoral work at the University of Illinois Urbana-Champaign, where he met George. At the time, George, all of eighty-four years of age, was Professor Emeritus at the University of Illinois. George has now passed away, but in his heyday he was a respected pioneer in radio astronomy. His website shows that he had worked on radio astronomy instrumentation, radio engineering, and radio direction finding; his areas of research included antennas for communication and wireless sensing, atmospheric and ionospheric measurements, coherent optics/imaging, electromagnetics and optics, and radio and optical wave propagation. Interestingly for us, George had also worked on physical acoustics (propagation, diffraction, absorption of sound, noise mitigation) and radio tracking and telemetry of wildlife. Indeed, he was in Panama assisting Martin with the creation of an animal tracking system on Barro Colorado Island.

One evening over cervezas and rum and cokes, we told George about the manakin wingsnap. Before we knew it, we had a collaborator, one with just the kind of scientific expertise and biological curiosity that could help us better understand manakins. The next field season, George returned to Panama with a sound engineer, and he had also recruited Daniel Bodony, professor of aerospace engineering at Illinois and an expert in aeroacoustic modeling, to assist. We thus set out collecting, from a variety of camera angles, high-speed video recordings of manakin males performing both wing- and rollsnaps. We set up an array of three microphones of known distance from the central area of a court where wingsnaps would originate. At about the same time, our collaborator Lainy Day (my previous postdoc who had become an associate professor at the University of Mississippi) had borrowed another high-speed camera, but one normally attached to a microscope with very bright lighting to record behavior of termites at speeds up to 60,000 frames per second. Lainy hoped to help us get really fast video recordings of manakin wingsnaps. In our case, the main limitation was lighting, but she was able to record a captive male wingsnapping at 2,000 frames per second! When examined, frame-by-frame recordings like these give individual frames separated by 0.5 milliseconds. When synchronized with audio, we had unprecedented visual resolution of wingsnap kinematics at the moment sound was produced.

We then set out to analyze all of the data we'd collected. An obvious finding from the synchronized audiovideo recordings was that the sound emanated from the wings at precisely the moment the wings collided—that is, within the half-millisecond level of resolution we could record.

These observations gave us several clues. First, both these video sequences and those of Bostwick and Prum show that the wings never move at supersonic speed and hence do not "break the sound barrier" to produce a whipcrack. Moreover, the videos show that the wings' primary feathers are still moving forward, and do not make contact when the snap is produced. This is in contrast to birds like pigeons, which make a clapping sound when they take off as their flight feathers strike each other—a sound quite different from the loud manakin snap. Thus, although Chapman doubted that the feathers could produce the snap, even his suggestion that sexually dimorphic flight feathers might be involved in producing the snap was shown to be incorrect. (In fact, we believe the sexually dimorphic feather structures had another purpose altogether.)

We could also reject the idea that the snap is like a human handclap, wherein the air is forcefully ejected from two broad surfaces brought rapidly together. This was not observed in the videos, and in any case, the wings simply could not have produced such a sound.

We concluded that the sound was produced by the percussion of the bird's outer wing bones—that is, the swollen tips of the bird's radii, located in the wrist. But how? This is where Daniel Bodony, the aerospace engineer, came in. Using the video images and sound analysis, Dan employed elementary acoustic theory to model several possible mechanisms of sound production. His findings, somewhat simplified, were as follows.

First, the sound could be produced as the air is pushed rapidly away from the contact point of the wrists. Dan's calculations showed that this would produce a sound with a maximum frequency around 400 Hz, considerably below what is in fact produced, so that hypothesis was rejected. A second possibility was that the wing movements produce an unsteady mass of air, as in the case of a jet engine or a rotating helicopter blade. Modeling of the wings' movements and accompanying pressure changes, however, showed that such a sound would carry its maximum energy at around 100 Hz—again, way below what is naturally heard.

The final possibility was that bone-on-bone contact at the moment the wings collide causes the bones to vibrate in such a way as to create the snapping sound. And sure enough, this appears to be precisely the case. The results of a model that accounts for the time scale of the impact and the density of avian bone show that such a contact would produce a sound in the range of 5–7 kHz, just what the bird in fact produces. We'd answered the question: the manakins throw their wings together powerfully enough to cause their wing bones to momentarily vibrate, emitting a 5 kHz wingsnap.

In my view, this is quite remarkable. Just think about what makes a bird a bird—namely, its wings and its ability to fly. But these little birds take their wings, these remarkable structures adapted for flight, and, throwing caution to the wind, say, "I can repeatedly slam my wing bones together, numerous times a day for months and years at a time, and I can still fly just fine. How about that?"

And the female manakins seem to say, "We're impressed!"

Quite frankly, so am I.

FIVE

Male Manakins Are Made to Snap

Ever accidentally smacked your elbow or shin on something solid? It hurts—a lot. So doesn't it seem that hitting two bones together repeatedly to make them produce loud sounds might also hurt? Even create damage? If you need your wings to fly, and you need to fly a lot, you might not want to risk damaging your wing bones.

But manakins seem to get by just fine. (Think woodpeckers, who bang their beaks into trees day in and day out with no apparent damage to their brains.) Indeed, manakins are incredibly long-lived animals given their small body size. Dave McDonald, a professor of biology at the University of Wyoming, spent years studying the spectacular blue-backed long-tailed manakin (*Chiroxiphia linearis*) in Costa Rica; one of the birds he banded was recaptured nineteen years later. And some golden-collared manakin males have been recovered from the same lek for up to ten years. That's a lot of wing-snapping! Clearly, the behavior is not damaging and, unless these birds are incredibly stoic, also not painful. What's going on?

Let's look at the bones of the forearm or wing for a second, bearing in mind that the skeleton of a bird wing, though strikingly sim-

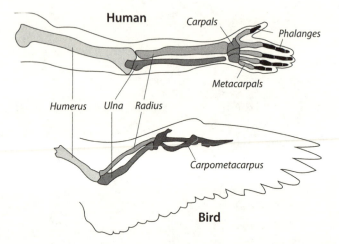

Fig. 5.1. The forelimb bones of a human arm and a typical bird wing.
Drawing by Bill Nelson.

ilar to that of the forearm of reptiles and mammals, including us, is also highly modified (fig. 5.1). For one thing, birds lack fingers and wrists; or rather, the bones that in us make up these appendages—the phalanges, metacarpals, and carpals—in birds are highly reduced and partially fused into what is known as the carpometacarpus. As a result, the bird's hand is quite unlike ours, functioning primarily as a target for the insertion of the primary flight feathers and thus adapting it for flight. Remnant minute carpals (not illustrated) then form the point of contact between the hand and the ends of the forearm bones—the radius and ulna. These bones are positioned to allow the hand to be extended and retracted to splay out the feathers, as when in flight, or to retract them next to the body, when at rest.

The bones of the forearm, the radius and ulna, connect the hand to the upper-arm humerus bone. These two bones are generally quite long, somewhat rounded in cross-section, and thickened at their distal end, where they contact one another and the carpals, as well as at their proximal end, or elbow, where, again, they contact one another and also the humerus. In many animals, the radius is the load-bearing bone of the arm, or foreleg, with the ulna running parallel to the radius, and sometimes even fused with it. In humans, the ulna is more fixed in its position, with the radius swiveling around in. As you turn your hand left and right and back again, the head of your radius, where your thumb emerges from the wrist, is moving over and around the head of the ulna, located beneath your hand.

The single, relatively long and robust humerus bone connects the elbow with the shoulder. Rotation of the whole arm occurs at the shoulder as the head of the humerus rotates within the shoulder joint, or rotator cuff, an incredibly complex and well-designed structure—except when it tears, as it does in many humans (especially sixty-plus-year-old people like me, and then it impedes arm movement, and it hurts).

These bones have numerous functions. They give the arm structure, providing things like muscles, cartilage, blood vessels, nerves, and lymph vessels a place to reside, and around which skin can adhere. The strength of these bones varies from one animal to another, depending on the weight they must bear. Most of the muscles of the forearm attach to these bones, so when they contract either the whole arm moves, around the shoulder joint, or its various parts move, around the elbow or wrist joints or, in the case of primates (including humans) in particular, in the hands and fingers. Thus, an additional need for strength on the part of the arm bones lies in the amount of force that contracting muscles place on them.

These bones in the wings of birds are generally quite similar to those of other tetrapod animals (those vertebrates with four limbs), but they are, of course, also adapted for flight. One key difference, therefore, is that wings are incredibly light, both in the sense of having hollow bones (in most bird species) and in the positioning and size of many of the muscles attaching to the wing and hand bones. These muscles are either decidedly reduced in birds, or they have been shifted to the bird's torso and away from the wings proper, reinforcing the bird's center of gravity beneath and between the wings. As a result, there are no heavy muscles out in the wings, but there are lengthy tendons that, in effect, connect the body of the bird to the hand. The same is true of birds' legs.

In one sense, therefore, birds' wings are rather fragile structures. Nevertheless, they can also keep heavy birds such as swans or condors up in the air for hours on end, they can withstand the profound air pressure of very fast flight, as in some falcons and swifts, and they can be banged together by manakin males to make loud snaps.

In the case of golden-collared manakins, it is the radius bone that seems to have undergone specializations to allow repeated radius-on-radius collisions, and to create and amplify the snapping sound itself. To investigate wing bones of manakins and some related species, we utilized micro CT (computerized tomography) imaging techniques, which allow three-dimensional structural images to be produced from numerous X-rays taken at different angles. (When humans are the object of such study, a regular CT machine is used; when you need to image something really small, like the radius of a small songbird, you use a micro CT machine!) This work was performed largely by Anthony (Tony) Friscia, a paleoanatomist at UCLA, and Matthew (Matt) Fuxjager, who started this work at UCLA as a postdoctoral fellow in my lab and is now an associate professor at Brown University. They made a number of key observations regarding adaptations of the manakin wing bones.

First, as shown in figure 5.2, the distal head of the radius (i.e. the end at the wrist joint) of the golden-collared manakin and related white-collared manakin (both in the genus *Manacus* and both performing loud wingsnaps) was seen to be enlarged relative to other manakin and nonmanakin species. Compare the radius of the red-capped manakin, a bird in the genus *Pipra* that does not produce similar wingsnaps. As this distal end is the site of the bone-on-bone collision that yields snaps, the enlargement seems to be a logical evolutionary step.

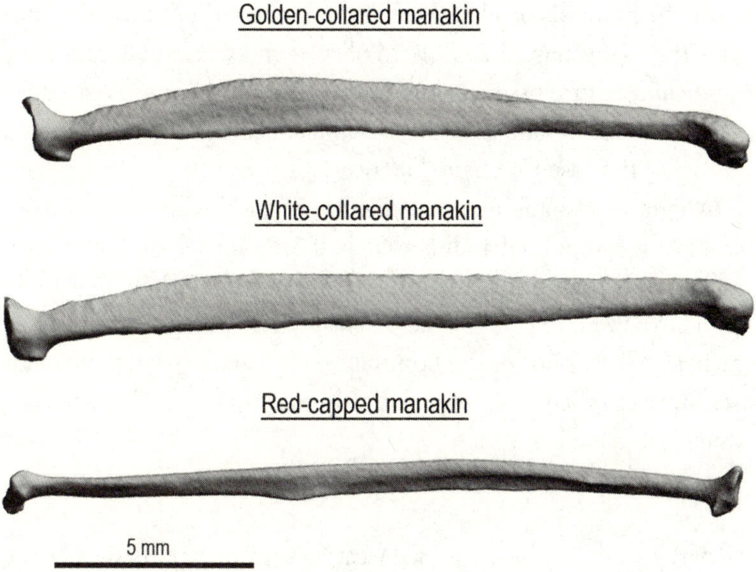

Golden-collared manakin

White-collared manakin

Red-capped manakin

5 mm

Fig. 5.2. The radius bone of three manakins, golden-collared, white-collared, and red-capped. Reproduced with permission from A. Friscia et al., "Adaptive evolution of the avian forearm skeleton to support acrobatic display behavior," *Journal of Morphology* 277 (2016): 766–775. Copyright © 1999–2022 John Wiley & Sons, Inc.

The second, and most conspicuous, feature of the *Manacus* radius is that, for most of its length, it is flattened like a board, unlike the radius of the red-capped manakin and most other birds. A typical bird's radius is more rounded in appearance. Notably, both male and female golden- and white-collared manakins possess this broadened radius.

A third remarkable observation was that the golden-collared man-akin radius is quite solid, in places completely solid. In contrast, the other two major forelimb bones, the humerus (connecting the shoulder and the elbow) and the ulna (the second bone connecting the elbow and the wrist), are both largely hollow. To be fair, the radii of all six birds we examined were a bit more ossified (solid) than either the humerus or ulna. Nevertheless, the radii of the two man-akins that wingsnap, the golden-collared and white-collared, were significantly more ossified than in the other birds, and the golden-collared manakin radius was in some places completely solid, which was never the case for any of the other birds.

In sum, wingsnapping manakins have typical wing bones except for having a modified radius that is flatter, more solid, and with thicker ends than in other birds. It is these adaptations that these small birds seem to need to make their loud snapping sounds. When the heavy distal ends of the bones make contact, a brief vibration is established in the bone, which vibrates the air that reaches the ears, which is detected as a snap.

As part of our studies, we often catch wild manakins by placing "mist" nets across narrow paths that we clear with machetes in the thick forest where the birds live. With the dark and often compli-cated background of underbrush, these nets of loosely woven thin black string are almost invisible to a flying bird. The nets are also soft and loose, so the birds are unharmed when they hit. Once in the net, their wings and legs and feathers do get tangled, but an experi-enced "mistnetter" can quickly determine from which side the bird entered the net and how it got tangled, and quickly free the bird.

That is to say, *most* birds are relatively easy to remove. Manakins, au contraire, are tough. They twist this way and that, and grasp the netting with their feet and hold on tight. One of the first tasks of the mistnetter, therefore, is to free the feet by grabbing and opening each toe individually. Some birds, like thrushes (the familiar Amer-ican robin, for example), will allow their toes to be opened with ease. Manakins, however, hold on for dear life, *and* they fight back. To free the legs, then, the mistnetter usually lays the bird on its back in the palm of the left hand, surrounding it with their fingers to pre-vent the wings from being damaged; they then use the right hand to free the toes. In so doing, it becomes clear that the manakin is a stout little bird, thick, muscular, and built to be strong. They are not mean birds, though. While some species of bird (especially those with thick beaks like seedeaters or parrots) will bite the hand that caught it, sometimes bringing blood, manakins resist not by biting but by squirming such that they become hard to hold.

Frank Chapman, when he studied manakins in the early 1930s, was so impressed by the build of the male manakin body that he sent several specimens to Percy Lowe, an expert anatomist, who commented: "On stripping off the skin of *Manacus vitellinus* the superficial muscles of the body are exposed, and it is at once apparent that certain muscles as compared with similar muscles of other species in this or neighbouring genera are considerably developed. This development takes the form of a remarkable hypertrophy of certain groups of muscles—of which I may mention the pectoral, gluteal . . . and scapular, and especially the pronators of the forearm. . . . [I]n the meantime I may state in general terms that a male *Manacus vitellinus* stripped of its skin looks like a well-fed little Quail, so pink and plump and translucent do the muscles of its pectoral scapular and gluteal regions appear. Indeed, as regards the latter they are so hypertrophied as to bear a strong resemblance to the buttocks of a naked human baby."[1]

Lowe's paper on manakin anatomy was published in the British ornithological journal *Ibis* in 1942 when World War II was raging. Apparently due to rationing, the journal used less ink and paper than before the war, resulting in images of lower quality. This is obvious in the drawing Lowe provided of the dorsal musculature of the male golden-collared manakin (fig. 5.3).

Strikingly, one muscle in particular stood out. Originally called the teres and infraspinatus, this muscle emerges from the back of the bird, somewhat above where shoulder blades sit on our backs. As pictured by Lowe, it is large and bulbous, very unlike the relatively flat backs of most birds. Recall that flighted birds have shifted the

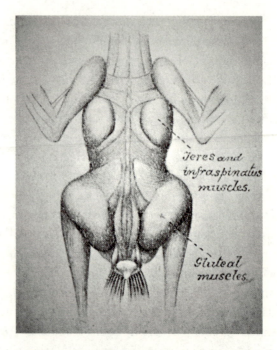

Teres and infraspinatus muscles.

Gluteal muscles.

Fig. 5.3. Percy Lowe's illustration of a skinned
male golden-collared manakin carcass showing the
enlarged teres and infraspinatus muscles (now called
the scapulohumeralis caudalis) and the enlarged
"baby buttocks" gluteals. Reproduced with permission from Percy R. Lowe, "The anatomy of Gould's
manakin (*Manacus vitellinus*) in relation to its
display, with special reference to an undescribed
pterylar tract and the attachments of the flexor carpi
ulnaris and flexor sublimis digitorum muscles to the
secondary wing-feathers," *Ibis* 84 (1942): 50–83.
Ibis © 1942 British Ornithologists' Union.

bulk of their wing musculature to the region beneath their wings,
on the body's ventral or front surface (i.e. their chest). Thus, a large
protruding muscle in the back of a small bird is unusual. Anatomists
now call this muscle the scapulohumeralis caudalis (or, for simplic-
ity, the SH muscle) because it is known to connect the scapula and

the humerus. When it contracts, rather than lifting the wing, it pulls it toward the midline.

In birds generally, the wings are lifted over the back by the ventral supracoracoideus (SC) muscles, and this is undoubtedly the case in manakins as well. So why is the SH muscle so large in golden-collared manakins? Could it be to help produce the wingsnaps so characteristic of this species? While this question has long intrigued us, it has not been an easy one to answer. But thanks to the tenacity of Matt Fuxjager, and some relatively sophisticated new technology, we can now say, without doubt, yes.

To determine if a muscle contraction is coincident with a particular motor movement (i.e. behavior), we can make electromyographic (EMG) recordings from discrete muscles across time as the behavior is performed. In relatively large animals performing relatively mundane behaviors, such data can be obtained relatively easily. But obtaining quality EMG recordings from small animals performing somewhat complex rapid movements is quite challenging.

As discussed previously, we know that SC muscles lift the wings, pectoralis muscles depress the wings, and SH muscles retract the wings. How do we actually know this, other than by simple analogy to the ways these muscles function in other vertebrates? We know because of elegant and pioneering studies by individuals such as Ken Dial, now professor emeritus at the University of Montana. Ken and his collaborators obtained EMG recordings from these muscles of birds in active flapping flight—no easy task.

First, pigeons were trained to fly down a hallway. Once the pigeons were at ease with this task, they were wired with electrodes inserted into specific muscles. As the birds flew down the hallway, and the wings lifted up and pulled back down, the EMG equipment would record when and how strongly each muscle contracted to produce these motions.

Some of these studies show that the principal ventral muscles, the SC and the pectoralis (PEC), contract in an exquisitely choreographed pattern. To obtain lift and to gain forward momentum, birds first raise and then lower their wings while also drawing them forward. As long as a bird is actively flying, as opposed to gliding or using air currents, the birds must repeat this motion, smoothly and effortlessly, time and time again. To raise the wings, the SC contracts. Before the wings reach their apex, the PEC contracts, apparently slowing the upward movement of the wings and then allowing for the wing depression as the SC contractions diminish. This produces an exceptionally smooth transition between the upstroke and the downstroke. The same set of contractions, in reverse, produces another smooth transition at the nadir of the wingstroke: that is, the SC begins contracting before the wings reach their lowest point, and the upstroke begins as the PEC contractions diminish. By this repeated cycle of contractions/relaxations, efficient flapping flight is accomplished. Just imagine how reckless flight would appear if each wing was slightly out of phase, or one or the other wing was raised a little too high or a little too low. But that doesn't happen, because the neuromuscular dance between the two SCs and the two PECs is impeccable—which we know thanks to EMG recordings.

We also collected EMG recordings of the SH muscle as the male golden-collared manakins performed wingsnaps (see plate 6). To promote wingsnap behavior, birds were first treated with testosterone to ensure the males were motivated to wingsnap. About two weeks later, when birds were snapping regularly, we gave them some light anesthesia and proceeded to put an electrode wire under the skin and into either the SC, the PEC, or the SH muscle. The wire was coiled so it would remain in place but move freely as the bird moved; it was then passed under the bird's thin skin to its back, where the wire emerged and was attached to a small (< 1 gram) radio transmitter held in place by adhesive. The next morning, with cameras and microphones ready, we turned the receiver on and waited to see if the birds would snap, and if so, whether the electrodes would become dislodged and the transmitter go flying. The birds did snap (well, some of them), and the electrodes and transmitters remained fixed in place (well, most of them). While we waited for the birds to snap, they also sometimes flapped their wings as they moved from perch to perch in their small cages. We recorded EMG data for these flaps as well, to compare with what happens during a snap. After obtaining recordings from several birds, we removed the electrodes and transmitters and released the birds back into the wild.

All of the telemetry data, as well as the acoustic data from the microphones, were digitized so we could synchronize and thus see precisely (within a 1-millisecond, i.e. 1/1000 of a second, timeframe) the process of contractions in relation to the snap itself—the moment when the wings collide at the top of the wing stroke. We could also see just how the same muscles contracted during a wing flap.

We identified three phases of the wingsnap: an ~80 msec period when the wing is elevated, the ~25 msec period when the wing is retracted and the snap produced, and a postsnap period, when the bird flutters the wings just a bit. It was quite obvious that the SC muscle contracts a small amount during the wing elevation phase, if not even a bit before. However, this contraction is not especially strong. In contrast, the SH muscle begins to contract during the wing elevation phase but strengthens significantly during the wing retraction phase, reaching a peak at the very moment the snap is produced. Not surprisingly, the PEC, as the primary depressor muscle, does not contract at all during either of these two periods, but does contract much later in the postsnap interval when the wings are lowered. Although we expected to see a stronger SC contraction when the manakin lifts his wing, the SH contraction was just as we expected: the powerful contraction of this muscle seems to be responsible for pulling the wings together with force, and it is this movement that creates the loud wingsnaps that we hear.

So the drawing that Percy Lowe made back in 1942 showing the bulbous SH muscle on the back of the bird was a true depiction of the muscle responsible for contracting to produce the behavior most associated with the golden-collared manakin, its firecracker-like wingsnap.

Scapulohumeralis caudalis is indeed a mouthful—such a long, complicated name for a relatively small, obscure muscle. But it is key to what makes a golden-collared manakin a golden-collared manakin. We suspected that the muscle was special from Percy Lowe's illustration, and we confirmed this by taking measurements of this muscle and by comparing male and female golden-collared manakins with each other and with male and female zebra finches. (The zebra finch is frequently used in comparative studies of small birds because it is one of the few passerine birds that thrives in captivity.) We found that overall the SH muscle was larger in adult male manakins than in adult female manakins or adult zebra finches of either sex. We also stained sections of the muscle and measured the diameter of individual muscle fibers (i.e. muscle cells) and found, again, that the SH fibers were larger in males than in females or in zebra finches. Finally, we obtained evidence that these fibers exhibited properties of "fast" muscles—we just didn't know how fast!

Skeletal muscle cells (fibers) are generally categorized as slow or fast. The biochemistry of slow fibers (sometimes called slow-twitch fibers) is such that they take longer to reach peak tension, but they also endure contractions for long periods of time. Fast fibers, in contrast, can twitch quickly, but they wear out more rapidly. Fast fibers come in two types, called fast glycolytic or fast oxidative; the former are faster but more prone to fatigue, while the latter, though still pretty fast, are resilient for longer periods of time. We found that the manakin SH muscle was a mixture of fast glycolytic and fast oxidative, but with the fast glycolytic predominating. Thus, the manakin SH had the potential to be quite fast and yet somewhat less prone to fatigue.

Of course, the real test lies in visualizing the SH muscle in action, and this was accomplished by Matt Fuxjager working with Professor Franz Goller at the University of Utah (now retired). Franz is an expert in muscle function in birds, studying especially the syrinx in songbirds, which produces the essential contractions that produce song. Matt, Franz, and their students set out to measure the contraction speeds of the PEC, the SC, and the SH in the golden-collared manakin as well as in several other manakin and non-manakin species. They were able to do this by using anesthetized wild birds and placing electrodes on the muscle of choice, which in turn was attached to a force transducer. The electrode provided rapid electrical stimulation of the muscle, and the force transducer measured the strength and timing of the resulting contractions. They found that "muscle twitch speed dynamics" were not particularly interesting in the PEC or the SC, being relatively similar across species and functioning in a range expected from previous work with other birds. The SH muscle, however, stood out as unique, but only in two species, the golden-collared manakin and the red-capped manakin (which also has fast wing movements in its courtship display). In these birds, the SH muscle is "superfast," contracting at roughly twice the speed required even for the bird to fly!

These results are amazing and indicate that the SH muscle is the fastest limb muscle of *any* vertebrate. Presumably, the biochemical and structural modifications of this muscle are required for the birds to perform their courtship displays, including the loud wingsnaps. As is discussed in a subsequent chapter, females prefer to mate with males that produce more wingsnaps, so female choice seems to have driven these birds to evolve superfast wings.

The observation that female choice creates in males a uniquely adapted muscle is exceptional. It is not necessarily surprising that males would have different muscles than females to support a true sexually dimorphic purpose. Many mammals, for instance, have specialized muscles for controlling the penis that are reduced or absent in females. Similarly, muscles in the necks or backs of males that carry large horns or antlers on their heads are more developed than those in females. But we might not expect that a bird's limb muscle would differ so profoundly between males and females when both need to fly. Could males suffer a cost by altering the properties of an important flight muscle? Conversely, could males gain an advantage by having one really fast muscle that, though developed for courtship, might also allow the bird to fly faster? And how does mate choice make just that one muscle fast, leaving the others unaffected?

All of these questions have a bearing on our understanding not only of sexual selection in general, but also of the ways this selection acts to create a distinguishing set of masculine traits, or phenotype. We believe that androgenic hormones are essential in this, something we explore in the second half of this book.

Two additional anatomical adaptations unique to golden-collared manakins deserve our attention, and these involve their feathers.

This species belongs to the genus *Manacus*, a group also called the "bearded" manakins. The name derives from the elongated feathers beneath the lower mandible of the male, which in the courtship display he extends outward, like a stiffened flag (gold in the case of *M. vitellinus*). Apparently, the males have skeletal adaptations coincident with these feather modifications that allow the beard to be held erect. As Percy Lowe noted in 1942, "[I]n comparison with other species of Pipridae to which I have had access[,] the muscles of the tongue, hyoid apparatus and glottis are rather conspicuously well developed, a condition of things doubtless correlated with the shooting forward, beyond the tip of the bill, of the bird's 'beard' when he is displaying."[2] The beard must be very important to the male for him to have expended so much evolutionary energy modifying his anatomy to enable such a distinctive behavior. Moreover, as Al Uy, a professor of biology at the University of Rochester, has shown, the bird's color keys to an appropriate forest background, which likewise contributes to female choice (i.e. the foliage around the leks of white-collared manakins in Costa Rica differs from that against which the golden-collared manakins of Panama display). It may be the beard itself (which is the same color as the collar) that females pay attention to. In any event, males go to great lengths to show it off.

Compared to female golden-collared manakins, males also possess modified primary and secondary wing feathers. One can see this clearly in photos provided by Frank Chapman in 1935 (fig. 5.4). Pictured on the right side of each image are the outermost elongate feathers of the manakin's wing, the primary feathers, so called because they arise from the hand of the bird and because they provide the power and lift for flight. In comparison to females, these feathers appear narrow and pointed in males. To the left of the primary feathers are secondary feathers that arise from the bird's forearm and are crucial for maintaining the integrity of the trailing edge of the wing. These feathers appear slightly curved to the left in males, with a thicker, more conspicuous shaft (or rachis) extending for the full lower half of each feather. After Chapman sent specimens to Percy Lowe, Lowe replied in a 1942 paper: "It seems to me that the sound described by Chapman as 'a low reedy whirr' is almost beyond dispute caused by these specialized primaries, as Chapman himself suggests. When, therefore, I come to the description of the feather muscles of the remiges I do not propose to say anything by way of description of the primary feather muscles. I am confirmed in this intention by the fact that the Hon. Anthony Chaplin tells me that *Manacus manacus*, which he has kept in captivity and which was in the habit of flying about his rooms, cannot help making a whirring sound every time it flies. It is as automatic as the whirr of an elephant hawk-moth."[3]

What is this "low reedy whirr" mentioned by Chapman and Mr. Chaplin? Unlike males, females fly about silently, whereas, within the

Fig. 5.4. Outstretched wings of a male (left) and a female (right) golden-collared manakin. Reproduced with permission from Frank M. Chapman, "The courtship of Gould's manakin (*Manacus vitellinus vitellinus*) on Barro Colorado Island, Canal Zone," *American Museum of Natural History Bulletin* 68 (1935): 472–521.

lek, males produce a whirring sound, ostensibly with their wings. Although the whirr presumably arises from the specialized wing feathers, the males appear able to turn the sound off. In contrast to Chaplin's claim, we have witnessed males' silent flight as they departed their lek in the direction of a fruiting tree; a few minutes later, perhaps after a meal, the males returned, again with no wing noise at all. This suggests that they can very accurately control the positions of their wing feathers to either make whirring noises or not.

Chapman had originally considered the possibility that these modified feathers gave rise to the wing snap. We now know that this is not true. If you clip the tips of these flight feathers, as was done by Lainy Day, males wing- and rollsnap just fine. We do not know if males produce the quack-like "grunt" of the grunt-jump display or

the reedy whirr if the feathers are clipped. However, we have never heard a female produce the grunt-jump or the reedy whirr, so we assume, along with Lowe, that these feather modifications in the male indeed produce the grunts and whirrs.

Why do males "whirr" when flying around the lek? We believe it is for arena (territory) defense, so other males know when an arena is occupied. Males are also attracted to the courting of other males, and the whirr sound of an approaching male may briefly distract a male in the midst of his dance. One way or the other, it seems to be an important feature of the male's behavioral repertoire, and a distinctive auditory feature of the entire golden-collared manakin lek.

SIX

Research in Field and Laboratory

I hope it is clear from my descriptions of all the complexities and wonderful beauty of the golden-collared manakin just how much I love them. You might then ask, how is it that, in doing research on them, I occasionally kill them? A very good question and one worthy of an inspirational answer. Our lives and those of other animals are intricately interwoven, providing much fodder for ethical considerations. Each of us brings our own unique assortment of experiences and understandings to the debate about the treatment of animals. We all possess our own natural levels of empathy, indifference, and machismo. Perhaps learning a bit more about me will help you see how these questions apply to our research program on the golden-collared manakin.

I do not know where it came from, but I have a strong love of nature in general. This love spans a huge swath of the intellectual and emotional space in my brain. The desire to understand something better is called curiosity, and I have it big-time about many things in nature, but especially about birds, manakins in particular.

It is often the case in biology that to learn more, we have to study specimens that are dead. As a consequence, our concepts regarding death and cadavers and dissection collide with those of life and wonder and love.

My life as a biologist, and the development of my perspectives about animals, started when I was quite young, and largely came from my early experiences as a fisherman and outdoorsman. Two memories stand out.

I grew up in Dallas, back when Dallas was still a big small town, not yet the metroplex it is today. Although I lived in a true neighborhood, with ranch-style houses separated by driveways and alleys, and yards and sidewalks and kids on bicycles, within just a few hundred yards of my house were small patches of riverine forest, a few small ponds, and one large undeveloped patch of prairie, crisscrossed by ephemeral "cricks," as we called them. On summer nights, our yard was filled with the flash of fireflies and the song of toads. After heavy rains, we sometimes found crawdads in puddles in the street. Occasionally horned toads ran across the driveway.

Many mornings during the summer, when I was about six, my mother would pack a small bag lunch for me, with a few extra slices of white bread wrapped in wax paper: my bait. She'd then walk me the few short blocks to Hillcrest Road, which, back then, with its two lanes and occasional car, was the nearest "busy" street (today it is six lanes wide and full of Texas-sized SUVs far exceeding the 40 mph speed limit). I'd examine my $5 Timex, and we'd coordinate my return back across the road at exactly 4 pm. After seeing me safely on my way, she'd go back home and I, my pole over my shoulder like a kid from Mayberry, would go fishing. At the pond, and by myself, I would find a spot and drop my line in the water. I caught mostly bluegill and tiny largemouth bass. I would stare into the water and watch the bluegills as they swarmed around my whitebread-ball bait. I tried to imagine what they were thinking. If

I hooked one, I would carefully remove the hook and place the fish back in the water, only to catch it again a few minutes later. I got to know each of the fish that grouped around my bait.

Once when I was fishing a group of older boys, nine or ten years old, came to fish at my spot. I watched. Upon hooking a bluegill, one that I most likely had previously caught myself, they would pull the fish from the water and, swinging the line over their heads, smash it onto the ground behind them, all the while laughing. They then left the fish flopping around until it died. I was frozen, saddened and angered by their crass actions. I wanted to defend the fish, but the boys were about twice my size and there were three of them. Even at that early age, I knew something was wrong: it was wasteful, savage, and senseless.

As I grew older, I'd go walking in the small patches of forest near my house. One day, I took along my older brother's discarded BB gun. In one oak and pecan grove, I came across a small bird poking around in the underbrush, probably some kind of sparrow, and I shot at it. Amazingly, I hit it, and killed it. In an instant, that little world that was alive just moments before, now was quiet. I picked up the lifeless little body, a drop of blood dripping from its beak, and felt devastated. I can still remember how still the woodlot seemed and how I had been the one to create that emptiness. That small bird was part of the whole life of the forest, and now the forest was a bit deadened. This incident has remained with me for more than fifty years as a profound emotional memory.

My early experiences went beyond just being part of the outdoors. Rarely, when fishing, I caught a fish big enough to keep. It is perplexing that I had such a different emotional response to catching a "big" fish, one that I took home, as opposed to a small fish, which I gently returned to the pond. In retrospect, it seems hypocritical. But hey, I was a young boy, and I was thrilled to catch a "big" fish. I was proud to bring it home to show off my accomplishment; perhaps we'd even eat it for dinner. Once I reached the house,

I would take the fish out back and clean it in the middle of our driveway. My mother was not pleased about this. But I remember that I was intrigued by the fish as I cleaned them. I would dissect all the parts to try to see how they all fit, what they might do. I'd inspect the stomachs to see what they'd eaten. I remember being amazed at how many organs could actually fit so perfectly in the fish's small belly. This dead and dissected fish, lying on our hot concrete driveway, was, to me, the final stage of the full sequence of bio-emotional events that began when I sat observing the small live fish swimming around my fishhook: from life to death and dissection.

As a youth, I did not appreciate that I was experiencing such a mixture of emotions, from joy to sadness, from wonder to thrill. They were all unfocused and unchallenged in my brain. For me, observing, catching, and then dissecting fish was equivalent to the developmental experiences of other boys my age with other hobbies, who were riding and then repairing ever-faster bikes, honing their passing and tackling skills, or shooting first skeet, then doves and ducks. Over the years, the emotional and intellectual states that fishing sparked in little Barney solidified, becoming organized and unified as a somewhat permanent feature of my character.

My early connection with fish gave way, for a few years, to an interest in whales and dolphins. While I was at college in Boston, however, the world of birds captured my soul. For me, college was a biological experience. I took every course on ecology, evolution, flora and fauna that I could. Boston was also an incredibly cold experience. One of my classes was a lab course in animal behavior. The instructor, Ben Dane, assigned us to go sit at Newburyport harbor (where the Merrimack River flows into the Atlantic Ocean, on the border with New Hampshire) and watch goldeneye ducks perform courtship behavior. While this might sound fun and easy, I can assure you it was far from that. These birds engage in courtship in February, when the temperatures in northern Massachusetts are in the single digits and the winds blow from the northwest. For the best views of the ducks, we sat in a spot where the wind blew over the water directly into our faces, unquestionably producing a wind chill well below 0 degrees Fahrenheit. This Texas boy was miserable. But at the same time, I was profoundly amazed. This bitterly cold harbor was absolutely full of birds: goldeneyes, buffleheads, mergansers, scoters, eiders, long-tailed and black ducks, and mallards, Canada and brant geese, some eight species of gulls, including one called the Iceland gull, two species of cormorants, an occasional bald eagle, goshawk, or rough-legged hawk, and even flocks of small rock sandpipers. And then there were the snowy owls. Once while watching several male goldeneyes head-bobbing for a single female, a fellow student yelled, "Hey look at that ice floe." Of course, seeing ice on a river was a big deal for me no matter what. This chunk, however, sported a snowy owl, and in its left talons was what appeared to be a bufflehead duck. My first snowy owl, and on an ice floe—and with prey. I was thrilled.

Once I walked half a mile across a frozen marsh toward a small juniper on which a pure white adult male owl perched. I would walk 30 meters or so, plant my spotting scope to get a better look, and then continue closer. The owl saw me but never moved. When finally I was within about 20 meters, and the owl's head filled my scope's eyepiece, he and I had a staring match. There we were alone, each regarding the other. That owl had no intention of giving up that perch. This was a transformational experience for me. Two unique minds, in a deeply cold and desolate spot, considering one another. After several minutes, I left the owl in peace. I would study birds for the rest of my life.

Encounters with nature that I experienced as a young man had their origins in my experiences as a boy, and they are alive and well in me now in my mid-sixties. I still love to go fishing, but instead of bluegill and little bass, I now surfcast for stripers and bluefish, at the same time keeping my eyes open for all the birds: combining the joy of fishing and the inspiration of the birds.

But I am not just a fisherman-birdwatcher. I am a professional biologist. I have three academic degrees in biology, and I am a professor (and past department chair) in UCLA's Department of Integrative Biology and Physiology. I have a lab filled with microscopes and chemicals and all sorts of laboratory equipment, small and large. I have offices and computers and students and postdocs. I also have a colony of approximately three hundred zebra finches housed in our Life Science Vivarium, comprising several large aviaries that enable the birds to fly about and interact socially. With plenty of seed, water, vitamins, perches, and playthings, the birds take care of themselves. They sing and court and mate and raise babies that grow up to sing and court and mate and raise babies. We observe these birds to see just how they behave and learn, interact and sing. Then, to fully understand these behavioral capabilities, which at their core depend on the bird's unique anatomical and physiological attributes, we sacrifice some individuals to study their brains, muscles, and hormones.

I also travel to Panama, where I do many of these same things with both wild and captive manakins. I have spent countless hours observing and listening to manakins at their leks. To get a better look, I have crawled on my belly to lie motionless within a meter of an arena where unsuspecting males performed. We often record the male's natural behavior with audio and video equipment. Sometimes, we manipulate the male's courting arena to test ideas about their responses. We also set nets to capture the birds. Sometimes they are just given an identifying leg band, or a small transmitter to record heart or muscle activity, and released. Occasionally they are bled so we can measure blood hormones, or implanted with a device that slowly releases a hormone or a hormone blocker. In some cases, the birds are taken into captivity, where they are tested in one way or another before later being released back where they were caught. In some cases, the birds are killed so we can study their brains or muscles.

The decision to take a bird into captivity or to kill it is never made lightly. We assess each lek and the number of males holding court. In order to maintain a lek's integrity, we never remove too many adult males. Some of the leks we have studied have been in existence for the twenty years we have been doing research in central Panama. If a decision is made to kill a bird, it is "sacked" as quickly and painlessly as possible, usually under anesthesia. Moreover, we strive to utilize as many tissues as possible from that one bird to minimize, overall, the numbers of birds we kill and to maximize what we can learn from each bird. We often contact other ornithologists to see if they can make use of any tissues we collect. We are required by both U.S. and Panamanian law to obtain permits to collect birds and ship tissues to North America. Further, we must obtain permission from both UCLA and the Smithsonian Tropical Research Institute to do any research with the birds, justifying everything from behavioral observations to the killing of individuals. There is tremendous oversight.

Make no mistake: we have strong feelings about what we are doing, and all of us find a balance between what we consider appropriate in order to gain knowledge and whatever harm we might be causing the population of birds in the area. Killing an animal is a solemn act that must be performed with respect. I would not allow a bird to be taken any other way.

But the work does not stop there. With each passing year, we reevaluate the population of manakins that live in and around our area of research. When we do an experiment comparing two or more groups, we need a sufficient sample size to enable rigorous statistical testing power. For example, if we are comparing males to females, or male golden-collared manakins to male red-capped manakins, or juvenile males to adult males, we usually include a minimum of three birds per group, and for some experiments as many as six. And even if our immediate interest has something to do with these birds' brains, we will also collect and freeze muscles, hearts, spinal cords, and other tissues for possible examination at a later time. Thus, each bird is used as fully as possible, with waste minimized.

If we feel that a lek has been impacted too heavily by our research, we turn to other leks. The leks that we have identified may be in relatively close proximity, within 500 meters or so, or they may be more distant, separated by tens of kilometers. If we need to kill twelve birds during one season, we might take individual males from twelve different leks or, if a lek is particularly large, up to two males at a time. Because females and juvenile males wander a great deal, we are able to collect these birds from many sites, not just adjacent to lekking adult males. Areas where trees are in fruit often provide good concentrations of birds. But again, we take only a few birds from each site.

Thanks to these efforts, we have seen little to no impact on the local population of manakins. They are still common birds around the town of Gamboa, where our research is based (see plates 7 and 8). Many leks are still active after years of experimental work, and although some have disappeared, it is unclear if our work was a cause or if other forms of human disturbance were the main factor. Our study sites are near the Panama Canal, and certainly a few leks have been destroyed when construction workers dumped huge volumes of rocks and soil dredged from the canal. In another instance, a local ecotourism hotel built a dirt road and a fake "Indian village" on and adjacent to one of the leks we had studied. While many of those males departed, a few stayed put and continued to perform at their arenas, despite now being less than a meter from the path and clearly visible to passersby. It became difficult for us to study these birds because our audio recordings of their wing- and rollsnaps now also included drumming and dance music emanating from the fake grass huts nearby. (An authentic indigenous village exists about 500 meters from this site; we have never heard this kind of drumming or music from there.)

All and all, we have been able to maintain a rigorous research program that has produced over forty scientific papers on the behavior and physiology of golden-collared manakins, and the birds are still there. We hope our work has gone further than just creating new knowledge that can be applied to our thinking about physiological properties that might advance medicine. Through our papers and talks at conferences, and the videos we use in our courses and that we share with other teachers, we hope that people will gain a greater appreciation for these remarkable birds and for the overarching need for large-scale animal and ecosystem conservation efforts. Finally, we also hope that our work has contributed concepts that can advance our understanding of animal physiology but also of evolutionary processes, including fundamental questions about mate choice and sexual selection.

SEVEN

Sexual Selection and Mate Choice

It is commonly believed that in the animal kingdom, and in other life-forms as well, males and females invest in the process of sexual reproduction in wholly different ways. By this I do not mean the act of coupling between males and females, nor the mechanism by which male gametes interact successfully with female gametes; rather, I mean the kinds of investments that males and females make toward their ultimate reproductive fitness.

We can most easily think about this by considering the general case of humans. An adult male makes up to 90 million gametes a day, in the form of sperm. It is widely accepted that an adult female makes no gametes at all after approximately midway through her own embryonic development. Even then, she makes only a total of about 7 million gametes, which are quickly surrounded with non-germ cells that form a protective cellular ball (called a follicle) around each individual gamete, or ovum. Despite this protection, many of these gametes die, such that a typical female has only a few hundred thousand ova once she passes through puberty and experiences menarche. These now come on line to be released, generally, one by one, at a special time called ovulation when the woman's body is prepared for successful fertilization, implantation, gestation, and ultimately, birth. Only one of the 90 million sperm that an adult male makes per day is needed to fertilize that one ovum made and nurtured for up to fifty years. Which individual sperm cell wins is largely a matter of chance.

This difference between males and females is one of investment, both in the generation and support of gametes and in terms of what occurs after the gametes are released. Adult males invest little in each individual sperm, but females invest a lot in each ovum. Moreover, given that female mammals (for the most part) experience internal gestation, the female, unlike the male, must invest a huge amount of her energy, time, organs, resources, and emotion in carrying the fetus, and then, because she is also the baby's food source, she supports its development for sometimes years to follow. Certainly, if a male is to reproduce successfully, he must assist his mate in some or much of this work. But he does not, by any stretch of the imagination, invest as much as the mom. Thus, the pre- and postcoital investment by males and females is a substantial basic difference in humans, and indeed in most other sexually reproducing organisms.

One potential outcome of such a system of reproduction is that males are free to attend to other nonreproductive activities. They are also free to engage in reproductive activities with other females, which can, in some cases, increase their overall reproductive success. Adult females, on the other hand, because they invest so much in each gamete, fetus, and baby (hatchling), place great emphasis on that one offspring, which ultimately will grow up to pass on her genes. For this reason, females tend to be very choosy about their mate, selecting only the best possible candidate(s) to fertilize their ova. Because males invest less, they need not be so choosy about which females they inseminate.

Such discrimination by individuals of the qualities of a mate is a form of selection with vast evolutionary consequences. Known as sexual selection, it was first described by Charles Darwin in 1871, a few years after he published his theory of evolution by natural selection.

Sexual selection differs from natural selection. From a highly simplified perspective, natural selection has to do with life and death and how many offspring one is able to produce, leading to subsequent generations. This form of selection includes avoiding predators, finding food, and being able to survive through especially tough times. Sexual selection also refers to the number of offspring produced, but the variables here have to do specifically with attractiveness to the opposite sex. You can live to a ripe old age, outsmarting predators, hunting and gathering with ease, successfully navigating times of drought or famine or disease, but if nobody chooses you as a mate, all is for naught.

Mate choice applies across sex. But continuing with the example started above, we can ask, just what are females looking for? This is a matter of great debate. Arguably, the male described above, with his many skills vis-à-vis the challenges of natural selection, would seem to be a good candidate for mating by a female. But how would she know this? Do males give off some signal regarding their survivability? And what if this male devotes too much of his time to survival—would he pay enough attention to his own offspring? A female might not choose such a male, instead preferring someone who will assist her with the many challenges of birthing (nesting), feeding, and raising a child. But again, how would a female know that the male will be faithful and helpful?

To fully appreciate how these ideas relate to manakins, we need to focus on three main concerns. First, a female might be looking for some kind of signal, or signals, that she can rely on to reveal a male's quality. The second, and related, concern is that males may themselves produce signals that provide information that is attractive to females. Third, we can actually view the first two ideas from the perspective of the female's sensory systems, her hearing or her eyesight, for example, as well as the male's physiological and anatomical attributes that match her most sharpened senses.

One can imagine a variety of signals that males might offer to convince a female that they are desirable for mating and fertilizing her precious ova. Perhaps not surprisingly, this area of study is highly prone to debate and speculation, but also to scientific observation and study. If a male does signal his qualifications to a female, how can she know if his signals are accurate and not "fake news"?

Some signals are hard to fake. Age, for example. In nature, survival is a great task: the individual must survive into and then past puberty. But, perhaps surprisingly, in nature, young adult males often fare poorly in the mating game compared to older, more experienced adult males. This may very well be true for manakins, so let's delve into this issue a bit further before returning to our consideration of manakin mate choice.

One can readily imagine that really old males must have considerable quality to have survived all of life's challenges. Some features of old age are hard to fake, but they may be hard to identify in a chance encounter. Males may want to advertise their age. Male, not female, mountain gorillas change from blackbacks to silverbacks at about 12 years of age, a clear signal of age. What causes age-related signals to develop? In many cases, we simply do not know. However, in one case we do know. In many human populations, older males lose hair on the tops of their heads. This particular form of hair loss, called alopecia, is induced by androgens; hence it is called androgenic alopecia. Thus, we have an age-related signal induced by a sex steroid. Such a relationship suggests that, at least in humans, androgenic control of balding has (or perhaps had) a connection to reproductive fitness. Signals of age within a reproductive context likely have considerable relevance to many animals, including, we think, manakins.

So, females might choose to mate with older males because they have proven they can survive a long time. But couldn't a younger male just fake being old? The female needs a signal that cannot be faked. Perhaps the female can set some bar for performance that cannot readily be performed by a younger male but is one that requires much practice? The task might be a complex motor task that is not required for any other purpose other than to signal to females that a lot of practice had occurred and so the individual male who performs well must be somewhat old. The task might not be physically demanding, but might require mental practice. For example, the male might need to demonstrate that he can solve a difficult problem that could only be accomplished after much trial-and-error learning. Or maybe there is no real task at all, but some other true signal of competence acquired with age. A male that has lived a long, healthy life might acquire resources that are displayed in horns or in feathers such that larger horns or feathers colored with specific pigments might signal prior successes over a relatively long lifespan. These might be attractive to females. Whether physical or mental or ornamental, these traits may function to signal age and females may prefer to mate with males that can prove they can stand strong against the non-reproductive challenges of nature.

Arguably, age could be irrelevant. Females might just want to mate with the most competent male. Imagine two males of equivalent age, but one male is a little smarter, a little more physically gifted, or just better at acquiring resources important to a female. Presumably, that male would receive more matings. Thus, independent of age, some "better" quality of a male might be what the females choose. The signal in question is not about age but about some male quality. Potentially, then, a really competent *and* older male would be at a terrific advantage.

One last important point to consider. This involves the general concept of "sender-receiver." Whatever signal is sent, it has to be received. In the cases we are discussing it means that sensory systems of the female must be tuned to receive the signals of age or competence sent by the males.

For example, if the best signal of male quality is some pigment in a feather, then the female visual system must be tuned to that signal. If, by chance, the very best signal is in the ultraviolet part of the visual spectrum, then her retina must have photoreceptors that respond to ultraviolet. Now, of course, we have a chicken-and-egg question. Which came first, the female's ability to perceive ultraviolet or the ability to acquire and insert ultraviolet-reflecting pigments in his feathers? Perhaps the chicken-and-egg mechanisms work in tandem over the course of countless generations to hone the photoreceptive properties of the retina and the male's talent for acquiring ultraviolet pigments. The bottom line is that both the males and the females are participants in this evolutionary process whereby female choice drives phenotypic traits of both the males and females of a species.

In all likelihood, Charles Darwin did not realize how much study and debate would arise from his descriptions of sexual selection. To this day, that debate still swirls within publications and symposia at meetings. I can hardly think of any area of biological science that is fraught with so many disparate viewpoints, with many scientists stubbornly defending their data and perspectives. I am quite sure that some of those scientists reading this section are bristling at my over-simplification and the angle I am taking with regard to manakins.

Let us also not forget that Darwin's thoughts about sexual selection in part arose from his fascination with birds, and not just all birds. As Darwin stated, "Secondary sexual characters are more diversified and conspicuous in birds. . . . The males sometimes pay their court by dancing, or by fantastic antics performed on the ground or in the air."[1] Following up on Darwin's observation, the study of manakin courtship can be seen as a wonderful opportunity for consideration of the processes and functions of sexual selection.

In considering sexual selection in manakins, our first question might be, just what is a female manakin looking for, or listening to, or sniffing at, or touching to gain knowledge about the males that come courting? In fact, we believe that females pay attention to *many* traits in a male, including his pigments, his arena, the noises he makes, his physical prowess, perhaps his smell, and his age—all a severe test of the male.

Let's take, for starters, his sounds (i.e. his auditory signals). It is the general consensus that one purpose of all the snapping produced on a male *Manacus* lek is to advertise a group of males to females at a distance in the forest. From the standpoint of the female, such a group is likely to contain at least one male that might be suitable to father her potential hatchlings. Some of our evidence for this comes from our large forest aviary, where we often observe green manakins (presumably females) approaching the perimeter and peering inside when it is occupied by displaying (snapping) birds. In the wild, too, we find that overall lek activity increases when suddenly one male's behavior abruptly increases in intensity. It seems that the first male to see an arriving female descends to his court and displays, an action acoustically picked up by the other males, who join in. As expected, we usually soon see a green bird scouting about in the lek, with each male either dropping down to his court and vigorously performing his jump-snap display or perching a few meters up and emitting a series of rollsnaps while anxiously scanning about, looking for the newly arrived female. Thus, acoustic signals are an undeniable feature of the male's courtship: males make unique sounds that are heard by females and which are downright attractive.

But sounds alone are not enough to satisfy a female as she inspects several noisy males. She needs visual stimulation as well to assist in her choice of a mate. Assume a female selects a male based on some quality of his courtship display. If so, then it follows that males must differ from one another, with some males being "better" and some being "worse." Because the display is performed so rapidly, it is difficult for observers like us to assess just what the male is doing. Moreover, it is very rare to see copulations, the unquestioned indicator of a female's choice. But with enough patience, they can be observed, and if you happen to have a high-speed camera at the right place and time, you can capture the brief romance and decipher just what is going on.

Thanks to my colleague Leo Fusani, we know that males do differ from one another in individual elements of their courtship acrobatics that visually stimulate the females.

A study using high-speed video was conducted during the height of the breeding season (March–April) as follows. After identifying males that we could videorecord with little disturbance, we observed the birds' behavior to determine the precise positioning of their courtship display. While some males jump between branches a few centimeters above the ground, others jump around a meter up, so the camera needed to be placed at just the right spot (4–5 meters from the center of the court), with just the right zoom and focus set, so that the full display could be captured. Then the males had to be caught and given a colored leg band for individual identification. Finally, the birds needed not only to return to their arena to display (which they all did), but also to ignore the presence of us, the beasts that had just caught, banded, and released them, as well as the newly installed tripod-mounted camera nearby. This took anywhere from two to five days, after which videotaping could begin.

Overall, about one hundred video sequences were captured from eleven different males at speeds of 125, 250, and 500 frames per second. Of these, some sixty sequences were of sufficient quality for analysis. The video files were transferred to a computer that ran software specially designed to allow frame-by-frame analysis of behavior with high temporal and visual resolution, so we could document the initiation and termination of any specific behaviors. The jump-snap display was the focus of the study, but that routine is composed of a number of discrete moves. We chose the following conspicuous behaviors for analysis: (1) *on perch:* amount of time spent perching on a sapling between two jumps; (2) *jump duration:* elapsed time between takeoff and landing the jump; (3) *jump speed:* duration of the jump relative to the distance between the two saplings; (4) *beard up:* time required for the bird to resume his statuesque posture with the erected beard at the end of the jump, from the moment of landing to the freezing of the posture; and (5) *snap frequency:* the number of wingsnaps per second during a single snap-jump display. From the multiple video sequences of these different males, we could compute an average value for each of the above behaviors. We could then ask questions regarding the variability of the behaviors of a given male, or whether the males differ from one another. Do some males routinely perform faster than others? Do some males perch longer between different behaviors? Do some males jump farther or faster?

Overall, the analysis showed that each individual male's movements varied by only a small degree from dance to dance—often mere milliseconds, suggesting remarkable neuromuscular control and consistency. When compared with other individual males, however, performances differed, in some cases quite a lot. For example, whereas one male stayed on the perch for less than half a second between jumps, another perched for over a second and a half, more than three times longer. Was the second male out of shape and more tired than the first? One male jumped across his arena at a speed of over 5 meters per second, while another's jump speed was about 1.5 meters per second, again more than a threefold difference. Did the first male have stronger legs than the second?

A particularly interesting behavioral difference was seen in the time it took the bird to raise his beard, the feathers under his lower mandible. Whereas one male could get his up (so to speak) in 42 milliseconds, another took 62 milliseconds, or nearly 50 percent longer. Now, 20 milliseconds may not seem like much to you and me. Indeed, our visual system doesn't even allow us to see an event lasting only 20 milliseconds. But female manakins seem quite able to resolve such a fast visual event, and it seems to have significant meaning.

It is this latter point that is crucial. The acrobatic routine of the male is an amazing visual signal to the female. Combined with the male's plumage and attributes of the arena, the female has much visual information to consider. Assuming they can see what our high-speed cameras see and can assess the differences that our computer programs allow us to decipher, females have an immense set of signals to consider, which, we assume, give them a solid basis for choosing males with the best genes to pass on to their offspring.

Bear in mind that the male must also see these movements, or he'd be performing blind. Apparently both males and females have extraordinary visual systems. From our perspective, it is simply re-markable. When the high-speed videos are played back in slow mo-tion, we witness just how beautifully precise are the movements of the male. It bears repeating that the males are reminiscent of human athletes, especially gymnasts. When jumping, they seem to throw themselves out with a heave, like a long-jumper taking off at the foot of a sand pit. The male manakin lands solid, again like a gym-nast at the end of a routine: legs together, no bounce, hands (wings) to the side. Even the slow ones produce an impressive performance. The fast males—well, they score tens all round!

With all of this new information in hand, we are still left with the question of just what all of this noise and acrobatics is signal-ing. What information is amassed by the female on which she bases her copulatory decision? This is an extremely difficult question to answer, but one that is amenable to testing. Do some males have superior neuromuscular capabilities as revealed by the speed and accuracy of their performance? In addition to admiring their beauty and skill, when watching male golden-collared manakins displaying, time after time, for a few hours each morning and afternoon, one

Plate 1. An adult male
golden-collared manakin.
Photo by the author.

Plate 2. An adult male
golden-collared manakin.
Photo by Lorie Humphrey.

Plate 3. A juvenile male
golden-collared manakin.
Photo by Ioana Chiver.

Plate 4. The forest in March in central Panama housing a
golden-collared manakin lek. Photo by the author.

Plate 5. A golden-collared manakin court in central Panama. Photo by the author.

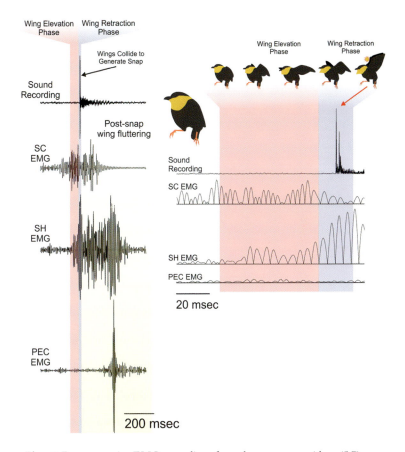

Plate 6. Representative EMG recordings from the supracoracoideus (SC), scapulohumeralis caudalis (SH), and pectoralis (PEC) muscles during the wing elevation (pink shading) and wing retraction (blue shading) phases of the wing-snap. The left-hand side of the figure shows the three muscle recordings (EMG traces) in sync with the sound recording of the wingsnap event. The right-hand side of the figure shows EMG traces magnified at the two behavioral phases. Above these signals is a schematic of the different wing movements manakins produce during the behavioral phases of the wingsnap. Reproduced with permission from M. J. Fuxjager et al., "Neuromuscular mechanisms of an elaborate wing display in the golden-collared manakin (*Manacus vitellinus*)," *Journal of Experimental Biology* 220 (2017): 4681–4688. Bird illustrations by Meredith Miles.

Plate 7. Dry-season forest of Soberanía National Park. Photo by the author.

Plate 8. The small town of Gamboa where many scientists live and work and where the Smithsonian Tropical Research Institute has created some of the finest facilities in the world for research in tropical biology. Photo by the author.

Plate 9. An adult male golden-collared manakin fitted with a miniature transmitter to record the bird's heart rate, with a blurry image of the author in the background. Photo by a member of the Schlinger manakin research team.

Plate 10. Gynandromorph zebra finch: pictures of the same bird with male plumage on its right side and female plumage on its left side. From R. J. Agate et al., "Neural, not gonadal, origin of brain sex differences in a gynandromorphic finch," *Proceedings of the National Academy of Sciences USA* 100 (2003): 4873–4878. Copyright (2003) National Academy of Sciences, U.S.A. Photos by Bob Agate and Art Arnold.

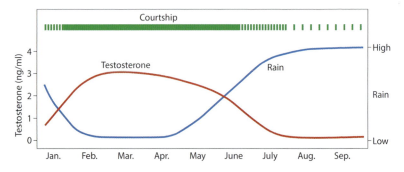

Plate 11. Seasonal cycle of male golden-collared manakin courtship, testosterone, and daily rainfall. Adapted from Barney A. Schlinger et al., "Hormones and the neuromuscular control of courtship in the golden-collared manakin (*Manacus vitellinus*)," *Frontiers in Neuroendocrinology* 34 (2013): 143–156, with permission from Elsevier. Drawing by Bill Nelson.

Plate 12. The unused Discovery Center aviary as we found it, with Matt Fuxjager as occupant. Photo by the author.

Plate 13. The Discovery Center aviary six months after the soil was deposited and saplings were planted in manakin-favorable patterns. Photo by the author.

Plate 14. A female manakin in the aviary feeding on local berries.
Photo by Ioana Chiver.

Plate 15. A view across the hybrid zone in western Panama. Photo by the author.

Plate 16. A view from a manakin lek within the hybrid zone of western Panama. Photo by the author.

does wonder if they get tired. In selecting a male for copulation, is a female impressed by the male's metabolic capability?

To assess whether manakins are taxed by all their exercise, we called on our colleague and friend Martin Wikelski, whose experience includes measuring bird heart rates using miniature telemetry systems. This procedure involves placing a tiny backpack on a bird which collects heart rates via small electrodes inserted under the thin manakin skin (see plate 9); a battery-powered transmitter then sends the signal to a handheld radio receiver. The entire apparatus weighs about 1 gram and falls off the bird after about five days, when the battery has gone dead.

Within a day or two after being fitted with the backpack, the males seemed not to notice their new gear and began performing courtship as if nothing had happened. We were able to collect the heart rates of birds over the course of full twenty-four-hour periods, including when the birds were performing the most physically intensive of their courtship routines. What we learned was outstanding.

First, we discovered that during the day, when flying about but not performing courtship, their hearts beat about 500–600 times per minute. This seems fast, but is in line with what is known for other small mammals and birds. At night and during their midday naps, rates of 300–400 beats per minute were observed. But when the males were performing rollsnaps and other courtship behaviors around their arenas, their heart rates shot up to well over 1,000 beats per minute, as fast as or faster than a tiny hummingbird's. While this remarkably fast rate only lasted for the ten seconds or so of a courtship dance, it represents one of the fastest heart rates ever measured, an indication that the manakin heart is an extraordinary structure indeed.

We learned something else as well. By monitoring the bird's heart rate, we could determine its overall energy expenditure. This involved calibrating heart rate with O_2 consumption and CO_2 production in the laboratory. We discovered that over the course of a day, the manakins seemed to expend no more energy than another tropical bird, the spotted antbird, which performs no obvious physically demanding courtship. These data suggest that manakins are well adapted for performing their elaborate courtship displays without needing to consume too many extra calories. Their bodies, and their hearts in particular, have evolved to give them the cardiovascular and metabolic profiles of an exceptional athlete. Perhaps the females have some sense of this when inspecting the male's performance.

It is important to point out that we do not know *why* males differ from one another. When human athletes compete, there may be various reasons for a specific outcome. Perhaps they differ in their natural abilities—that is, in their genetic predisposition to perform in a certain way. Perhaps the winner practiced more than the loser, or had a better coach, or performed more bench presses, or used some kind of illicit drug (such as a steroid hormone). We might consider whether one athlete was confident and calm, the other nervous and hesitant. Maybe the victor has competed in more such events, or is helped by the "home field advantage." In other words, genetic, physiological, experiential, and psychological elements may all combine to create the athletic differences we observe.

Is the same true of male manakins, and does a female sense all of this in selecting her romantic partner?

PART II

Endocrinology and Behavior

EIGHT

Behavioral Endocrinology and Vertebrate Sex Differences

As we continue our exploration of manakin behavioral biology, we will increasingly see that hormones play a key role in creating much of the magic that we are lucky enough to witness. To fully appreciate how hormones shape manakin behavior, we will need a basic understanding of endocrinology. This will tell us not only about manakins (and other birds), but also about ourselves, for we humans, along with virtually all vertebrate organisms, possess the very same endocrine systems. Thus, the study of manakin behavioral endocrinology teaches us a great deal about ourselves, while our knowledge of human endocrinology guides our understanding of manakins. Let us start, then, with some basic physiological principles and build these into an appreciation of the overarching roles that hormones play in shaping brains and behaviors, until we arrive at the dancing manakin himself, and a female who is keeping watch.

Animal bodies are basically sacs filled with fluids and organs. In a vertebrate animal, an internal skeleton gives that bag shape. But what makes the bag do things? What impels it to move, to feed, to drink, to sleep, to desire sex? All of these behaviors (and more) are guided by the central nervous system, which controls the contraction of muscles with exquisite precision. Coordinated muscle contractions are stimulated by neural impulses that arise in motor neurons, most located in the spinal cord but with some present in the hindbrain. Contractions are also guided by sensory systems: located in and around muscles are receptors that can sense the position or stretch of a muscle and send that information in the way of neural signals back to the central nervous system via their cell bodies located in dorsal root ganglia that run alongside the spinal column. Although the spinal column has a considerable degree of independence in coordinating behavior, the primary site where behavioral output is activated lies within the brain itself. Here premotor circuits fire to activate or inhibit motor neurons, leading to behavior. While the brain performs an extensive number of duties, the control of behavior is a core function, and one that can be powerfully impacted by hormones.

Now, consider a simple organism, say something very small and wormlike that likely existed many hundreds of million years ago. Such an animal may have performed the same set of behaviors all the time, such as moving constantly through the soil, ingesting whatever lay in its path, and releasing self-fertilized eggs as it went. From birth (or hatch) to death, there was behavioral output, but this simple worm lacked the need, or the capacity, for change.

Most organisms, however, rely on behavioral change. Imagine for a moment that you are a slightly more complex organism than the little worm just described. Your movement might carry you into an undesirable location, such as a place that is too dry or too hot; now you need to stop moving forward, you need to turn around and move elsewhere. Hopefully this redirected movement will carry you to a desirable location, with a rich supply of food, where you want to stop and mingle about. These are changes in behavior. They require sensory information about the presence of moisture, heat, or food, which then needs to feed back to activate the turning muscles or to instruct the movement muscles that it is time to slow down or stop.

Upon entering a place rich in food, you may determine that there are sufficient nutrients not just to keep you alive but to allow for the manufacture of eggs. Thus, you might not only be inclined to stay put, but you might be motivated to reproduce. Now you need a signal to tell the ovary to begin making those eggs and releasing them. If you are a sexually reproducing organism, as most are, then a male with testes would be instructed to make sperm and the males and females would receive new instructions to find each other and engage in reproductive (copulatory or spawning) behaviors. In vertebrate organisms, hormones created in the brain and pituitary gland activate the gonads to make gametes. The gonads, in turn, create hormones that circulate back to the brain to say they are ripe and ready for use: it is time to mate. The actual behaviors of finding a mate and then engaging in sex-specific actions to achieve fertilization can be very complex. But they evolved long ago and are all driven by hormones.

A hormone is a particular kind of molecule that is released into the fluids of the body (hemolymph in invertebrates, blood in vertebrates) and then travels from one tissue to another, ultimately imparting some effect on a target tissue. In the example above, the brain secretes hormones that act on pituitary targets, the pituitary secretes hormones that act on gonadal targets, and the gonads secrete hormones that act on numerous targets, such as the internal and external genitalia, but also the brain. What makes one tissue a target and not another depends on whether cells of that tissue respond specifically to that hormonal signal by expressing one or more kinds of receptors for that molecule. Once hormone bound, the receptors trigger a sequence of events within the target cell, such as changing its expression of genes, or changing its intracellular concentration of other molecules or ions, like calcium, that in turn activate or inactivate intracellular proteins to change cell, and thus tissue, properties. Organs can be made up of many such target cells, so the whole organ may become stimulated and active.

Neurons communicate with one another and with muscles, and they do so by also releasing molecules, neurotransmitters or neuromodulators. These chemical signals do some of the same things as hormones, but more typically they act by directly impacting the electrical excitability of membranes of other neurons or of muscles. A change in electrical properties of neurons can travel along nerve fibers, often a long way, to produce its effects. Motor neurons innervate muscle fibers, where they release a compound called acetylcholine, which increases intracellular calcium concentrations, causing the muscle to contract. Organs like the gonads are not muscles, so generally there is no need to have nerves from the central nervous system (CNS) reaching out to them. Instead, as described previously, the CNS primarily uses hormones as the mode of communication to the gonads, and as we will see, the gonads use hormones to converse back with the brain.

The principal hormones used by the gonads to communicate to tissues throughout the body are steroids. These fascinating and powerfully acting molecules are largely produced by two sets of organs: the gonads, where steroids participate in organizing animal anatomy and physiology and behavior appropriate for purposes of reproduction, and the adrenal glands, which daily regulate numerous bodily functions, but which also specifically modify many tissues in response to stress. Other tissues can synthesize small amounts of steroids, but we will restrict our focus to the hormones involved in reproduction. Appropriately enough, these hormones are the sex steroids. These sex steroids are crucial players in the manakin story, so we will devote a good bit of time to considering what makes these molecules such powerful biological signals.

■ ■ ■

Most of us grimace when we hear the word *cholesterol*, given its strong association with heart disease. Its bad reputation aside, cholesterol is in fact a crucial substance for all the cells in our bodies. It is also *the* source of our sex hormones progesterone, testosterone, and estradiol, thanks to a conversion process mediated by enzymes.

Think of a molecule of cholesterol as a collection of carbon atoms arranged in rings, with some carbon side-chains of various sizes sticking off the rings this way and that. Some of these carbons have oxygen or hydrogen atoms attached as well. To make a hormonal steroid from cholesterol, a particularly long side-chain of carbons must first be removed from one side of the cholesterol molecule. This reaction, called side-chain cleavage, leaves behind a backbone of carbon rings with some smaller side-chains—a new molecule called pregnenolone, a steroid. This step is key, because cholesterol likes to nestle into the membrane of cells, in part because of the presence of the long side-chain. Its removal allows the remaining steroidal backbone to pass freely through membranes. This capacity makes steroids exceptional signaling molecules, because once released into the blood they can essentially go wherever the fluid takes them and then enter virtually any cell in the body with ease. If the target cell has the right receptor for that particular steroid, a message can be delivered and the target cell will transform in some way in response. The ability to cross cell membranes and communicate information was already important for multicellular organisms hundreds of millions of years ago, and this steroidal signaling has only been improved upon since.

The steroid pregnenolone can do little on its own. However, an additional reaction can convert pregnenolone into the sex steroid progesterone, a progestin, an especially important female hormone. Progesterone may be the body's signal of choice. Alternatively, an additional few reactions might act to turn progesterone into the hormone testosterone, an androgen, an especially important hormone, often in males, or a final enzymatic step might convert testosterone into estradiol, an estrogen. Yes, you heard me right: the hormone that we typically think of as the quintessential female hormone, estradiol, is derived directly from testosterone, the hormone we think of as quintessentially male. The close connection between the two is important and leads to some biosexual surprises.

The enzymes that convert cholesterol into sex steroids are largely packaged together in specific cells in the male testes and in the female ovaries. In the testes, these cells are called the Leydig cells. The testosterone thus created readily exits the Leydig cell and diffuses through the testis to impact the process of making mature sperm; it also diffuses into blood vessels, where it is carried by the bloodstream to the rest of the body. The properties of testosterone that allow it to cross membranes so readily also mean it isn't terribly happy in the watery medium of blood. The presence of numerous proteins in blood, including some that specifically bind and carry testosterone, makes their voyage through the bloodstream easier and can facilitate their delivery to target tissues throughout the body.

A similar process occurs in females. Progesterone and estradiol are produced by two cell types in the ovary, the granulosa and theca cells, which are components of the ovarian follicles that house egg cells, or ova. A vascular supply to these follicles removes progesterone and estradiol (and even some testosterone) for passage through the bloodstream of female vertebrates. Once that blood enters the small capillaries in a target tissue, the sex steroids can diffuse out of the bloodstream and then across the outer membranes of that tissue's cells. The message has now been delivered and the next phase of hormone action begins.

What I have described here applies essentially to all vertebrates, from elephants to humans to manakins. The timing of hormone secretion and the actions on target tissues are a bit different in males and females of different species, but the sex steroid physiology is fundamentally the same.

■ ■ ■

Steroid hormones in general, but especially the sex steroids, can have a powerful influence on target cells, effectively changing who they are and what they do. They do so, however, only by binding to specific receptors. Think of it like a lock and key, where the hormone is the key and the receptor the lock. Each kind of receptor has a specific hormone that binds to it, and vice versa. This is true, for example, of the sex steroids progesterone, testosterone, and estradiol, which bind to intracellular receptors. In addition, there can be receptors that reside on the cell's outer membrane, where they can influence properties of these outer membranes as well as some intracellular properties. Such receptors can induce rapid changes in the membrane's electrical properties, producing powerful effects on behavior. Biologists are quite familiar with the receptors for different categories of steroids. Given their overall importance, they are very well studied, understood to tie to the proper function of a vast number of cells in the vertebrate body.

It is also the case that some hormones are more potent than others, like one key that opens a lock smoothly whereas another key has to be jiggled relentlessly before the lock accedes. Consider, for example, estradiol and estrone. Both are estrogens (they have the same overall shape and number of carbon atoms), but estradiol binds potently to estrogen receptors, whereas estrone binds poorly. Thus estradiol has potent cellular actions, whereas estrone usually does not.

What happens when steroids bind to receptors located inside of cells? While this gets somewhat complex, a primary event is that the cell undergoes a change in its expression of genes. The phenotype (the appearance and function) of each cell is guided by the way that cell uses its DNA. Although nearly every cell in your body contains essentially the exact same DNA, these cells differ from one another in that they use different suites of this DNA. Recall that

DNA is composed of immensely long chains of smaller molecules, or nucleotides, with some of the chains of DNA called genes. The activation of each gene involves first making a unique set of RNAs that are, in turn, converted into a unique complement of proteins. It is these proteins that determine what the cell is and does. Note that despite all cells having roughly the same DNA, not all genes are *expressed* in all cells, and even if the same gene is expressed in two different cells, the intensity of that expression may differ. Thus cells ultimately differ in the kinds and amounts of mRNAs, hence the proteins, that they possess. If you can change which genes are being converted into proteins, and regulate by how much, you have a powerful tool for directing cells along specific functional pathways. These differences in gene expression are accomplished by DNA that lies between genes that code for proteins. These regulatory lengths of DNA, lying upstream of protein-coding genes, are thus of profound importance. They are also the sites where steroids and their receptors bind to control gene expression and the functional properties of individual cells.

So what kinds of effects do sex steroids have on cells in various body tissues? Well, this is a bit like asking what things in your house you can use a screwdriver on. Many items in your house have screws that can be adjusted, causing both big effects and small. Steroids are like that. There are many tissues in your body that respond to sex steroids (i.e. have screws); some respond with dramatic changes, while others are just tweaked a little.

Let me give an example of a tissue that responds to a steroid in a major way. It's one I use in my classes. First, I ask my students, "How many of you had an egg for breakfast sometime this past week?" When I get a good show of hands I say, "Well, you may think you ate an egg this week, but you're wrong. What you ate is actually a package of estrogen-dependent gene products." This usually brings a collective grossed-out look, which subsides (sometimes) as I continue the story.

■ ■ ■

The eggs we eat are formed in the ovaries and oviducts of chickens and are composed almost entirely of proteins, lipids, some water, and various ions and other special organic and inorganic molecules, most of which are, one way or another, created or collected by the hormone estradiol acting on estrogen receptors located in the female's ovary, liver, oviduct, and bones. Estradiol binds to estrogen receptors in liver cells to increase the expression of genes that synthesize lipoproteins, which, secreted in the bloodstream, are taken up by the ovary and deposited into the ovum to make the mustard-yellow egg yolk. After this ovum is released from the ovary (ovulated), it travels down the oviduct. Cells lining the oviduct possess estrogen receptors, and upon binding to these receptors, estradiol profoundly upregulates expression of the gene that codes for the protein called ovalbumin. These oviduct cells then secrete this ovalbumin onto membranes surrounding the egg yolk, and it becomes a dominant component of the egg white. Finally, a shell is laid down on top of the egg yolk and egg white: here again, estradiol acting on the bones of the chicken causes them to release calcium, which is deposited by the bird's uterus onto the egg as its shell. Thus, an egg is the product of lots of estrogen action on many tissues. When you consider that a chicken can lay an egg every day for days in a row, and you consider how much fat and protein and calcium is involved, it is a rather amazing manufacturing process, all under the control of the circulating hormone estradiol.

This example serves to illustrate how one steroidal signal, estradiol, can impact a diversity of tissues, instructing them to do vastly different tasks. In the case of the profuse production of ovalbumin by otherwise relatively quiescent oviduct cells, the change in cell function is profound.

But what does all this have to do with the behavior of male manakins?

Let's consider a similar set of processes that occurs in vertebrate animals when testosterone is secreted into the bloodstream by the testes and that blood then bathes the brain with its androgen-enriched cocktail. Much as estradiol affects many organs of the female body, testosterone can act on various parts of the male brain; the effects may be relatively minor, but in some brain areas they are profound. These brain regions are occupied by neurons that express receptors for steroids, just like the cells of the liver or oviduct. And like those cells in the periphery, when a hormone binds to its receptors in neurons, there is a change in gene expression, the neuronal proteins are altered, and the overall chemistry and structure of the neuron changes. This can in turn produce a change in the connectivity and/or firing of a particular neural circuit—which becomes evident to us as changes in behavior. Sometimes these behavioral transformations are extraordinary, such as when a peaceful wintering songbird suddenly becomes an aggressive vocalist come springtime. It is also why male manakins dance.

Before we delve into the mysteries of hormonal effects on brain and behavior, let us consider a few broad, sometimes overlapping, and often controversial concepts regarding sex steroids and reproductive phenotypes. In virtually all vertebrates, males and females exhibit sets of morphological and behavioral traits that are appropriate for each sex. We generally label these sex forms as "male" and "female," and we apply the general term *phenotype* in this reproductive context. Thus, we consider together those anatomical features, physiological processes, and behaviors that seem to characterize that sex as being either "masculine" or "feminine." However, as we shall see, those perceived phenotypes can be deceiving, even in manakins.

When we think of sex from the standpoint of gender, especially in humans, more sex forms can be identified. Lesbian, gay, transgender, bisexual, etc. are different expressions based largely, but not entirely, on behavioral phenotypes. In adult birds, males and females are often anatomically, physiologically, and behaviorally quite different. In golden-collared manakins, for example, adult males bear a conspicuous plumage of bright golden-yellows, stark blacks, and subtle greens, and their elaborate physical courtship is visually stunning and aurally loud and attracts attention. Adult females, in contrast, are solidly subtle green; their plumage is so similar to the colors of the surrounding forest that they flit about mostly unseen, like small ghosts or sprites. Along with these behavioral and plumage phenotypes, adult males are physiologically and anatomically capable of inseminating females, and females are physiologically and anatomically able to be inseminated, to have internal fertilization, and to make an egg and lay it. Added to the behavioral repertoire separating males and females is that only female manakins build nests, incubate eggs, and tend to hatchlings, and probably help fledglings gain independence. Males are too busy maintaining their courts and displaying to participate in parental care. Thus, there are numerous clearly distinguishable differences between adult male and female golden-collared manakins. The two sexes do share one colorful trait in common, however: they both have bright red legs, unlike other species of manakins (or most other bird species) living in the same forests.

This seemingly simple scenario has one complication, though. As it turns out, young males are visually indistinguishable from females. Why young males possess the same subtle green plumage as females is a matter of debate. Perhaps they are avoiding competition with adult males for resources. Perhaps by pretending to be female they can sneak up on unwary true females and sneak copulations. Perhaps the disguise allows them to get a close look at adult males' dance routines. Perhaps the drab plumage allows them time to acquire the skills needed to avoid predators, a skill required when sporting bright yellow feathers. Indeed, there may well be various advantages to delaying growing an adult's plumage for a year or so.

However you look at it, there is a phase in the male's life when he appears to be female.

The adult male plumage is a clear masculine sexual trait. We are most familiar with our own human sex-specific characteristics. Some we call primary, some we call secondary, and others we are reluctant to label at all. Primary sex characteristics include the gonads and genitalia. Males have testes, a penis and scrotum, and a number of glands and tubes that connect these organs. Females have ovaries, labia, a vagina, clitoris, cervix, and uterus, and fallopian tubes. These are the sex-specific parts of our primary sexual phenotype.

Birds have many of the same features, though with modifications. Both sexes have a single opening, called a cloaca, that they touch to one another when copulating to transfer sperm from male to female. Birds typically lack a penis, and the male's testes are retained within the body cavity (external testes and penis would create a lot of drag in flight). Females, as we have seen, have modified their "fallopian tube" and uterus to form an oviduct and shell gland, for purposes of laying eggs. The wonderful mammalian uterus that undergoes extensive hormone-dependent modifications when an egg (ovum) is fertilized to become the nurturing, placenta-lined environment for internal gestation is absent in birds. Although bats manage it, flight would be difficult for a bird trying to carry a mammalian-like uterus, never mind a placenta and fetus.

The process whereby males and females develop along different trajectories is called sexual differentiation. In birds and mammals, and a few other (but not all) vertebrates, the sexes develop uniquely because they carry different sets of genetic material. In most mammals, females carry two copies of the X chromosome, whereas males carry one X and one Y chromosome. A gene on the Y chromosome, but not on the X, is turned on during an early stage of embryonic life to instruct a collection of pregonadal cells to become testes. In the absence of this gene, the pregonadal cells will spontaneously develop into ovaries.

The embryonic mammalian testes then start doing something quite relevant to our discussion: they begin secreting hormones. One of these hormones is testosterone, and when this hormone is present during very specific periods of fetal life, the tubes, glands, and genitalia appropriate for male reproduction are formed. Another hormone helps ensure that female structures do not develop. In females, which lack the testicular hormones, all of the tubes, organs, and genitalia develop properly to enable female reproduction.

In birds, this process of sexual differentiation occurs much as it does in most mammals, but with a twist. In birds the sex chromosomes are called Z and W, with males carrying ZZ and females ZW. Beyond that, the instructions that lead to the formation of an ovary vs. a testes are not fully understood. There is a gene on the Z chromosome that, when present as two copies (i.e. ZZ), causes testes to develop. By contrast, the instructions for ovarian development are probably not on the W chromosome but reside somewhere else in the bird's genome. Presumably, the genes that encode ovarian development are able to be turned on when the instructions for testicular formation are insufficient—that is, when the bird has only one Z chromosome. Thus, a little male manakin, tucked snug in his egg in a small round nest of soft plant material perched in a small sapling a meter or so above the ground, amid magnificent rainforest, naturally develops a pair of testes, while his sister in the adjacent egg develops ovaries in part because she will not grow testes. There is one additional trick. Because, as we have seen, birds do everything they can to reduce weight, and because a female can only carry one developing egg at a time—if, that is, she still wants to fly—she only needs (and wants) one ovary. So, shortly after the two ovaries begin to develop, a stop signal occurs in one of them, virtually always the one on the right. This ovary disintegrates, and the primordial germ cells that it held "jump ship" and cross over to safely enter the security of the remaining left ovary. Thus, all adult flying birds have a single ovary, and it is almost always on the left side.

All of the phenotypic characteristics so far described are fundamental sex characteristics. But there are more. In many vertebrate animals, some sex differences are created within the central nervous system, in the spinal cord and the brain. These often lead to differences in adult behavior and physiology, sometimes subtle, sometimes striking. For example, in many vertebrates neural circuits in the brain guide the specific copulatory behavior that matches the anatomical sex form. So males have circuits that make them attracted to females and able to engage in copulation, which often involves penile insertion into the vagina and ejaculation. Females, on the other hand, have neural circuits that guide typical female copulatory behaviors, which attract males and then allow those males to engage in penile insertion until ejaculation occurs. So, like the anatomical structures themselves—the penis in males, the vagina in females—the circuits in the brain controlling copulation are different in males and females.

Note that even in species lacking an external penis, like birds, males and females still perform precopulatory behaviors that signal a readiness for mating (or spawning). In birds, these behaviors may lead to a cloacal kiss, enabling contact between sperm and eggs and fertilization, and this "kiss" involves sex-specific postures to be successful.

How do such sex differences in the brain arise? Well, in much the same way as for the other primary sex characteristics. In male mammals, the testicular hormone testosterone travels to the brain and guides the growth or death of the circuits that allow the male copulatory circuits, but not female circuits, to be built; in females, the female circuits grow naturally. Similar neural circuits in birds develop naturally in males but are prevented from developing in females, being demasculinized by hormones secreted by the developing ovary of females.

These traits that arise developmentally are generally permanent, persisting in the animal for life. These hormonal effects that produce the sexual phenotype are thus called organizational, and can be differentiated from effects that are more transient in nature, which are often termed activational. For example, in seasonally breeding animals we witness aggressive and reproductive behaviors only during their breeding season, when their gonadal hormones are secreted. The same behaviors are absent, or greatly reduced, during the non-breeding times of the year because the gonads secrete little, if any, sex steroid outside of the breeding season. So the behavior is turned on or off depending on the presence of the hormone. Although they are often more obvious in other animals, humans demonstrate similar activational hormone effects beginning at the stage we call puberty.

Puberty is an extraordinary transformation. We are like two completely different organisms before and after puberty. Before puberty, male and female humans appear rather similar—that is, if you do not look at the genitalia and if you ignore culturally or socially based clothing and hair styles. There is considerable overlap in behavioral norms between boys and girls.

Puberty is the process where the somewhat quiescent reproductive system becomes active. The gonads that have been patiently waiting—the testes in their little sacs or the ovaries nestled within the body's abdominal connective tissue—come to life because a physiological and anatomical transformation occurs in the brain. This brain makeover causes a large release of a neurohormone called GnRH (gonadotropin-releasing hormone), which travels directly to the pituitary gland and triggers cells there to secrete two additional hormones, FSH (follicle-stimulating hormone) and LH (luteinizing hormone). FSH's primary role is to tell the gonads to wake up and grow and to begin making functional and releasable gametes, while LH tells the gonads to begin making steroid hormones. Those hormones not only enable the maturation of sperm and eggs, but also spread around the body to say, "Hey, now that we can make sperm and eggs, let's get all the rest of the machinery up and running to get them together."

This requires not only that the gonads release gametes properly, but also that the tubes and glands that transport sperm and eggs in males and females respectively are put to use. The way they are put to use is by having males motivated to copulate with females, and females motivated to copulate with males, and each sex must be able to perform the appropriate acts of copulation. The steroidal signals from the gonads travel to tissues in the body to ensure that the sperm and eggs can pass safely through their body's tubule systems, but also act on the vertebrate brain to respond to sensory signals of the opposite sex, thereby energizing individuals to copulate. In males this involves activating motor behaviors such as mounting, intromission, and ejaculation. In females it can involve activating approach and solicitation behaviors and, ultimately, the execution of physical postures that enable the male to intromit.

Before puberty, I thought girls had "cooties," yuck. After puberty, whatever girls had was something I badly wanted and would do just about anything to get. My reproductive system was awakened. My testes began secreting testosterone that grew the tubule systems required for the movement of sperm. Testosterone also activated neural circuits that were the basis of sexual desire and behavior. On top of this, however, were a diversity of other transformations that, though not absolutely required for successful reproduction, increased the odds that this might occur. These we call secondary sex characteristics, and in humans they arise during puberty.

With puberty human males, now men, have more body muscle mass, grow a little taller, grow significant body hair, especially on the face, and deposit minimal adipose tissue (fat), acquiring what we call a more masculine appearance. Not only are men interested in females, but they may become more aggressive or engage in other new behaviors in an effort, presumably, to dominate other males and to also attract females. These behaviors are not absolutely required for mating, but they may result in the increased likelihood that mating will occur. Behaviors like these are also secondary sex characteristics arising, it seems, from neural systems activated by testosterone. Females (now women), meanwhile, develop a unique shape to their pelvis and, unlike males generally, lay down adipose tissue under their skin, on their hips and buttocks, and on their chests in the form of breasts. Women also become interested in males and may engage in new behaviors to attract them. Sex, something that neither males nor females thought much about before puberty, becomes a powerful motivating force and a somewhat pervasive feature of the adult phenotype.

Successful reproduction does not mean that you are more likely to have sex, though that can enter in. By successful reproduction we mean that you are likely to pass on your genes to another generation, which in turn become successfully reproducing individuals themselves. It can take a long time to grow physically, psychologically, and cognitively into an adult that is ready to engage in the profoundly complex process of being a mature, sexually reproducing organism. In humans, this process takes ten to fifteen years or more, and can differ slightly between males and females. So, unlike other vertebrates, humans have a lengthy childhood. But when puberty occurs, with all of the elaboration of secondary sex characteristics and behavioral changes, it is a time for great celebration (mine involved an awkward Bar Mitzvah), if not also a time for a little trouble.

Of course, just because an individual is able to function as a sexually reproducing organism (i.e. an adult), it doesn't mean that success is immediately guaranteed. This can take many more years to achieve. In manakins, as in humans, considerable time after puberty may be required before a male in particular becomes a suitable mate or mating partner. This postpubertal development will be a topic of a later discussion.

For many vertebrates, the developmental process we experience as puberty can occur yearly. The sex behaviors and, in some cases, aspects of the anatomical and morphological sexual phenotype are only present at restricted periods of time when steroidal hormones are present. They are absent when the hormones are gone. Their transient appearance when the hormone is present implies that the hormone has activated the physical trait or the behavior. Thus, during the nonbreeding season when it is of no advantage, or perhaps even disadvantageous, to be sexual, the gonads secrete little, if any, sex steroid and the behaviors are not performed. When conditions for reproduction are appropriate, then sex steroids are produced and the sex behaviors are activated.

Once we humans have passed through puberty we remain, for the most part, ready and able to reproduce for much of our lives, females until menopause, males even longer. Even in humans, however, the drive to reproduce is not linear, and there are periods when the reproductive drive is stronger and weaker. Women go through periods when reproductive drive is reduced or enhanced, such as during or after pregnancy or in phases across the menstrual cycle. In many animals, the need to produce young during specific seasons that maximize their growth and survival means that periods of sexual drive are dramatically turned on and off. It is when sex behaviors are turned on in a seasonal breeding animal that the activational effects of hormones become readily apparent. The males of many bird species, for example, form flocks in winter, but once springtime comes they become territorial and aggressive, singing or crowing or, as with many manakin species, dancing to signal their presence and sexual interests.

As important as gonadal hormones are, they are not the whole story when it comes to phenotypic sex differences, and this may be especially important in birds. One kind of bird in particular offers us an extraordinary example where hormones seem to have little if any involvement in the acquisition of sexual phenotype; these strange birds are called gynandromorphs, a true natural oddity. What makes them so astounding is that they are male on one half of their body and female on the other half, and that division runs right down the middle of the bird's body. In a bird where the males and females have different plumage or body size, the effect is striking. The internet is now full of pictures of naturally occurring gynandromorphs that show up at people's feeders, like a North American cardinal with brilliant red plumage on one side and dull yellowish-pink plumage on the other. There is also the gynandromorph "chicken," where the rooster half is profoundly larger than the hen half so the bird stands with a striking lean to one side.

Because hormones go everywhere in an animal's body via the bloodstream, clearly hormones cannot be responsible for creating the bilateral sexual phenotypes of gynandromorphs. Studies from the laboratory of Art Arnold at UCLA have provided a partial explanation. In fact, gynandromorphs are just what they seem to be: genetically male on one half and genetically female on the other half. We are familiar with the notion in humans that females possess two X chromosomes (XX) and males have one X and one Y (XY). As described previously, birds also have sex chromosomes, but they are designated differently such that males are ZZ and females are ZW. In the case of one gynandromorphic zebra finch that Arnold's group studied, half of the bird was ZZ and the other half ZW.

What made this zebra finch so special (at least to scientists interested in such things) was not its remarkable bilateral plumage per se (see plate 10). Rather, it was the differences observed in its brain, which precisely matched the plumage difference: the bird's brain was mostly male-like on its right side and mostly female-like on its left, with the two halves pretty much divided down the middle. Arnold's group examined the brain because, as described in chapter 3, songbirds possess a well-defined, though complex, neural circuitry that controls song learning and performance (see fig. 3.1). In zebra finches, adult males sing but adult females do not, and as it turns out, several of the nuclei in the brain controlling song are larger in males than in females. When the zebra finch gynandromorph's brain was examined, sure enough, the song nuclei on the right side were larger than on the left.

Arnold's group delved further by examining the expression of genes on the bird's Z and W chromosomes. (Remember, since males are ZZ and females are ZW, all the cells in the male body should have two copies of any ZZ gene, while the cells in the female body should have one copy each of a Z gene and a W gene.) Images of the gynandromorph brain where expression of these genes can be visualized show what is going on (fig. 8.1). In these photomicrographs, the brain is viewed as if the bird is facing the reader, such that the bird's left side is on the right and the right side is on the left. Compare these images with the color photographs in plate 10 showing the bird's plumage, which is male-like on its right side and female-like on its left side. Examination of a W chromosome gene (ASW) shows it to be vastly more abundant on the left (dark staining of the top figure), with virtually no presence on the right side. Thus, the cells on the left side of the bird's brain carried the W chromosome, suggesting that they possessed the female-like ZW chromosome combo. Most cells on the right appeared to lack a W chromosome, suggesting they must be ZZ. This conclusion was confirmed when expression of a Z gene (PKCI-Z) was examined (note that there is nothing really special about these genes ASW and PKCI-Z other than their positions on the W and Z chromosomes, respectively). As expected, there was some expression of the Z gene on both sides, but the amount on the male/right side was about twice that on the female/left side. This is exactly what one would see if there was one copy of the Z gene in ZW cells and two copies in ZZ cells. Thus, the brain and presumably the rest of the gynandromorph's body had the male ZZ chromosome complement on its right side and female ZW complement on its left.

This work on gynandromorphs forces us to ask whether some

Right   Left

A

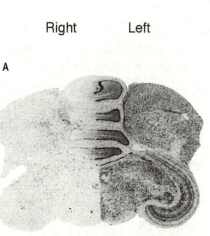

asw

B

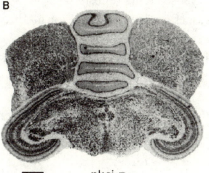

pkci-**z**

Fig. 8.1. Sexually dimorphic gene expression in sections through the brain of a gynandromorph zebra finch (viewed as if the bird is facing the reader). Reproduced with permission from R. J. Agate et al., "Neural, not gonadal, origin of brain sex differences in a gynandromorphic finch," *Proceedings of the National Academy of Sciences* 100 (2003): 4873–4878. Copyright 2003 National Academy of Sciences, U.S.A.

sex-specific phenotypes we observe in nature arise from the activational effects of gonadal sex hormones or from the expression of sex chromosome genes, or perhaps some combination of both processes. In the case of Arnold's zebra finch, the focus was on the brain, and structures in the brain were either masculine or feminine based on patterns of *expression* of sex chromosome genes. But we also know that sex hormones do create sexual phenotypes, so we are left with the tantalizing question of just what mechanism leads to the masculine or feminine traits we observe. When you see a sexually dimorphic bird fly by and you immediately recognize its plumage as male, is it because male genes in the bird's skin induced the deposition of masculine feather pigments or was the bird's skin exposed to testosterone or estradiol when the feathers were growing during molt?

This question is especially relevant to golden-collared manakins, where adult males and females differ profoundly in their plumage, but juvenile males look like adult females. Is the growth of the adult plumage dependent on exposure to hormones or is it a constitutive trait of the cells in the feather follicles? Do juvenile males turn off male sex chromosome genes in their skin until ready to grow their conspicuous golden and black feathering, or are they female green because their testes just aren't yet secreting testosterone?

In summary, whereas hormones secreted from the gonads direct the development and function of *some* of the body's sex-specific anatomical, physiological, and behavioral traits, there are other characteristics where hormones have no impact; rather, those features arise because of the expression of genes on the sex chromosomes of the bird's cells. We explore these ideas in the next chapter, where we consider what drives males to create courts, dance wildly, and rollsnap noisily.

NINE

Hormones Control Behavior in Manakins

On a typical Panamanian day in early January, the heavy rains of November and December having subsided, the dry season sets in. The burdensome heavy humidity lessens its grip, the sun becomes more prominent. The forest begins to change. Birds are singing, the cicadas are buzzing, a few frogs might still be calling, and fresh, dry breezes from the southwest rake the treetops. Leaves rustle and fall. New colors appear as trees begin to flower in bright yellow and magenta. And suddenly, the sounds of snaps emerge from within the forests. Manakin courtship is coming alive.

Adult male manakins that have been quietly traipsing about looking for berries, or that have departed these forests looking elsewhere for food, are suddenly back at their leks. They may have used the same court for many prior years, perhaps as long as a decade, disappearing for the few months of the heavy rainy season, but they return on schedule as the dry season commences. We do not know where they all go; we just know that they come back to the same spot—and we think we know why: testosterone.

I, together with my colleagues Lainy Day, Leo Fusani, and Martin Wikelski, have been intrigued by manakin courtship and the mechanisms that turn it on and off, and we have used several approaches to investigate whether testosterone activates male manakin courtship. The most straightforward way to assess this question is to measure the amount of the hormone in blood, the prediction being that adult males actively performing their courtship routines have higher levels than do birds that are not performing courtship or making much noise, the quieter birds being females and juvenile males, as well as males during the noncourtship season—that is, during the wet season, from June to December.

Indeed, this seems to be the case, as we found when we captured manakins of both sexes and across different ages at different times of the year. To collect a little blood from manakins, you first need to catch the birds. Recall that wild manakins are captured using a mistnet, which is virtually invisible against the dark backdrop of the shadowy forest. The birds flying about in their leks hit the net, which is softly cushioned with pockets of netting, and they lie there until removed by the investigator. Blood can be drawn from a small bird like a manakin by piercing a vein in the wing and allowing a few drops of blood to flow into a tube. Only a small volume is collected in part because only a small amount is needed, but also because birds are exceptionally good at coagulating their blood after an injury. The experimenter has to work fast. Once blood is obtained, a piece of tissue paper is held over the wound to ensure blood flow has stopped, and the bird is then released. Males, in particular, patiently wait out the brief procedure and, upon release, fly right back to branches above their court, look around, and start roll-snapping and displaying as if nothing at all just happened.

Once collected in the field, the blood sample is kept cold, on a slurry of wet ice, and returned to the lab. All of our work on golden-collared manakins has been performed at facilities run by the Smithsonian Tropical Research Institute, usually in a small room beneath one of the many duplex houses in the village of Gamboa built originally for workers on the Panama Canal. Inside such a "lab" room we keep a small centrifuge that we use to separate the red and white blood cells from the liquid blood plasma. The plasma is then frozen on dry ice (not always easy to obtain in Panama) and moved to a freezer for storage. At the end of a field season, all the samples are returned to a research lab Stateside for analysis of the hormone levels in the blood.

I say this matter-of-factly, but it is actually quite amazing that we can measure hormones in blood. One of the things that makes hormones so interesting is that they can have powerful effects even though they circulate in minute quantities. For testosterone, biologically active levels in birds are generally about 1–3 nanograms per milliliter (ng/ml), or 10^{-9} grams (one billionth of a gram) of molecules of testosterone per one milliliter of blood. Let me illustrate what this means. A teaspoon of sugar weighs about 9 grams. Mix that in a cup of coffee: you can imagine that the sugar imparts some sweetness. Then dilute a teaspoon of that coffee in a quart (or liter) of water. Do you think you could still taste any sweetness? Probably not. Now, take a teaspoon of this new mixture and dilute that in another quart of water, and repeat two more times. In the end, one teaspoon of the final diluted solution would have only a few nanograms of sugar molecules left, an incredibly small amount.

If you didn't know how much sugar was in the final teaspoon, how would you measure it? Fortunately for endocrinologists, researchers back in the 1960s and early 1970s figured out a method: radioimmunoassay, or RIA. This process involves mixing a sample, such as a small amount of blood, with two things: a bit of radioactive hormone, like testosterone, and a mixture of antibodies that specifically bind to testosterone and to nothing else. Recall that you have antibodies floating in your bloodstream all the time that can bind to nasty things that might enter your blood, like viruses or proteins on the outer membranes of bacteria. This is one way your body's immune system protects you from infection. The antibodies that recognize testosterone are often made in an animal like a rabbit; this is done by injecting the rabbit with a modified version of testosterone (it has to be modified because the rabbit does not make antibodies to its own testosterone) and then letting the rabbit naturally make antibodies to the modified hormone. Blood can then be collected from that rabbit and purified so that you have concentrated antibody to the modified testosterone but which also binds to natural testosterone. This antibody also binds to testosterone with a radioactive tag on it.

Back in the lab, when everything is mixed together—that is, the bird's plasma, the radioactive testosterone, and the antibody—the antibody can bind either to the radioactive testosterone or to the unknown amount of testosterone in the plasma sample. The more of the unknown amount of testosterone there is, the less is available to bind to the radioactive form. Thus one can "measure" the amount of the unknown by determining the amount of radioactivity that is bound to antibody: less radioactivity equals more unknown and more radioactivity means less unknown. There are a variety of ways to measure radioactivity, but do not fear. The kind of radioactivity used in these experiments is very weak and safe. Moreover, very sensitive devices can measure a single radioactive emission, so only a minute amount of radioactive testosterone is used in these kinds of experiments.

You may wonder why I've gone off on this biochemical tangent when what you want to hear about is manakins. First, I hope you appreciate that to ask a question about hormones in a wild animal is not such an easy task. In addition, it is simply hard to overestimate the importance of this procedure, not just for measuring testosterone in manakins, but for just about every test your physician performs on the blood sample collected during your annual checkup. The report you receive telling you how much cholesterol is in your blood, for example, relies on this or a similar procedure. The scientist who invented it, Rosalyn Yalow, was awarded the Nobel Prize for her work.[1] Nowadays, the procedure has been refined so that radioactivity is no longer needed; instead, simple enzyme-based reactions reflect the amount of hormone present by causing a color change in samples, which can be measured by a small affordable machine. One of these procedures is called ELISA, short for enzyme-linked immunosorbent assay.

The ability to measure hormones in wild animals greatly improved our opportunities to understand animal behavior. An aptly named scientist, John Wingfield (distinguished professor emeritus at the University of California, Davis), perfected the procedures for measuring hormones in small wild birds, opening the floodgates for ornithologists, ethologists, and behavioral endocrinologists to investigate whether hormones activated particular behaviors in their bird of interest.

For our studies of manakins, at the end of a season of research in the field we ship all of the frozen plasma samples back to our lab at UCLA where they are immediately stored in a freezer at –80°C. When ready to perform the RIA or ELISA, we thaw the samples, knowing that our birds are likely still dancing about in their Panamanian rainforest home. To make sure the measures are accurate, we run a number of samples of real or artificial blood that are blank or that include known amounts of hormone. In the end, we can tell if the collected blood has any hormones in it, and if so, just how much. Generally what we have found is that females of any age, as well as juvenile males and adult males during the very wet (nonbreeding) season, have virtually undetectable levels of testosterone in their blood. By contrast, adult males caught from their leks during the dry season generally have measurable levels of testosterone in blood, some as high as 6 ng/ml, which is, relatively speaking, quite a lot; the average for males caught in mid-January, shortly after they have arrived at their courts after their wet season hiatus, is 1–3 ng/ml (a level similar to that seen in many adult men).

Elevated levels of testosterone in males during the breeding season are consistent with the conclusion that testosterone activates the males' dancing behavior and their wingsnapping. This seasonal cycle of rainfall, circulating testosterone, and increased sex behavior is illustrated in plate 11.

Although testosterone and behavior are correlated, this relationship does not prove that behavior is causally linked to testosterone. One way to explore this is by means of "extirpation/replacement," in the words of endocrinologists. In the case of male manakins, one would want to rid the birds of the source of the hormone (i.e. the testes) and administer the hormone (i.e. testosterone) separately as a controlled dose. There is a relatively long history of the use of such procedures by, and even on, humans. Humans have long known that castration of domestic male animals made them less wild and easier to keep in captivity. Sometimes, they even taste better. Humans also know that castration at a young age makes men unmotivated (or unable) to have sex with women, hence the creation of eunuchs to guard royal harems. Humans understood this with regard to features of puberty as well, and would castrate boy choristers to prevent the adolescent deepening of their voices so they retained their angelic tone.

Interestingly, the first true experimental demonstration of the extirpation/replacement paradigm was performed on, you guessed it, a bird. In that experiment, published in 1849, Arnold Adolph Berthold surgically removed the testes of young roosters, and noted not only that they lost or failed to grow their masculine secondary sex characteristics (plumage, combs, and wattles) but also that they became nicer. Berthold also took a group of males that had their testes removed (capons) and then reinserted the testes of other males. Sure enough, these birds developed their masculine phenotype, including showing an increase in aggressive behavior and interest in sex. The simplest conclusion was that the testes were somehow involved in creating the male phenotype, with a more important conclusion being that the testes released some kind of factor that was responsible for masculinity.

Probably directly or indirectly inspired by this experiment, two of the greatest physiologists of all time, Sir William M. Bayliss and Ernest H. Starling, identified in 1902 a specific blood-borne factor produced by the stomach that influenced GI tract function. Called "secretin," this was the first confirmed hormone. It established the whole vast field of endocrinology, which now studies hundreds of hormones from dozens of organ sources that perform innumerable functions throughout the bodies of all multicellular organisms.

As you might imagine, it is hard to castrate a small wild bird like a manakin. So we have done the next best thing: we study them at the time of year, the wet season, when their testes naturally make little, if any, testosterone. During much of the wet season, especially near the start, adult males are still seen hanging around their courtship arenas and still performing some of their courtship behaviors. In addition to the adult males, there is also an influx of green birds, young males, some of which also begin displaying, though with much less vigor and skill. All of these birds have very low or undetectable levels of testosterone in blood. What happens if we give them more? This is analogous to an extirpation/replacement experiment, but instead of castration we are taking advantage of a seasonally induced loss of testosterone secretion, and giving exogenous testosterone back to them. We usually do this by a quick procedure that involves slipping a relatively small implant under the bird's skin. The implant, a thin plastic tube sealed at both ends, is filled with testosterone crystals, and the silica-based plastic is permeable to small molecules like testosterone. Once implanted in the bird, the testosterone slowly leaks out at a steady rate, so that after a day or so, the blood testosterone levels resemble those of breeding adult males, and they remain elevated for a month or more. The effect of this testosterone is striking.

Some of the birds received the same plastic implants, but they were empty. These individuals served as controls for the stress caused by the handling of the birds and the minor surgery to provide the implant. They behaved pretty much as they did before receiving the surgery; that is, they hung around their arenas but did very little if any courtship behavior. By contrast, those birds given testosterone performed dramatically more courtship behaviors: about fourfold more rollsnaps, sixfold more courtship displays, eightfold more wingsnaps, and many more calls of all sorts.

Because we were unsure how many males would stick around after being given an implant, we also kept a few in captivity, in small cages suitable for keeping finches at home. These birds, too, increased their performance of courtship behavior, generally by a wide margin. To control for nonspecific effects of our treatments, we also recorded several nonreproductive behaviors, such as how much they hopped and fluttered about in the cages. Here we saw no difference between birds treated with empty implants and those with implants filled with testosterone, suggesting that the hormone had no negative impact. What testosterone did do was specifically increase the suite of behaviors that we associated with courtship and/or the aggressiveness that males show to hold a display court for themselves.

Significantly, even males with little or no testosterone performed some of the courtship behaviors, albeit at a fairly low level. In other words, they had at least some of the physical capacity to do so, though they seemed unmotivated. As we have seen, an animal is no animal at all if it is not motivated to behave; rather, it is just a bag of organs and fluids. So what testosterone does is stimulate motivation to perform a set of behavioral routines.

We often think of collections of individual entities, whether it be people in societies or genes within genomes, as occurring in networks, with a few important hubs that serve to control the larger groups. Here, we can think of testosterone as stimulating a hub that then functions to increase behaviors connected to that hub—as if there is a center in the brain, a sort of manakin acrobatic courtship routine motivation center, upon which testosterone acts to promote expression of courtship behavior. If that is the case, where is that center and how does it connect to all the nerves, muscles, and sensory systems that allow male manakins to dance with such exquisite precision and speed?

Earlier we discussed sex differences in behavior and wondered whether females could have any capacity or desire to perform male courtship displays. We know that females have low levels of testosterone in their blood, whereas adult males have high levels. What if we gave testosterone to adult female manakins—would that make them act like males? Or will they never perform male courtship behaviors, no matter how much testosterone they are given, because they are, in fact, inherently different from the males?

To explore these questions more thoroughly, let us return to a consideration of singing behavior in songbirds. Recall that singing is controlled by a complex interconnected set of neural circuits. In adult zebra finches, males sing, but females do not, even with a dose of testosterone, and the adult female's neural song system is less well developed than the male's, and remains poorly developed in the presence of testosterone. Yet, consider another species, the canary. In canaries, adult females generally do not sing, and their neural song system is relatively poorly developed. But give the adult females some testosterone and the females will sing and their song system will grow to appear much more like that of the adult male canary. Thus, in the case of zebra finches the sexual dimorphism of the brain when it comes to singing behavior is rigidly established early in life, whereas in the case of canaries, this sexual dimorphism is not rigid, but is controlled by the presence or absence of testosterone secreted by the adult gonads. In other words, the brains of adult male and female zebra finches are fundamentally different, with the male having and the female lacking the basic "hub" site where hormones act to motivate and physically control the movements of the syringeal muscles that produce song. In the canary brain, in contrast, that "hub" site exists in both sexes; the only difference lies in which hormones are circulating, male or female.

Applying this perspective to manakins leads us to ask, do adult females possess the neural and muscular machinery to perform courtship but lack the testosterone to activate these circuits? Or do females lack the male's neural and/or muscular machinery to perform male courtship, which would mean that no amount of extra testosterone will stimulate females to perform male-like behaviors?

These questions have been difficult to assess experimentally. When Lainy Day was working in my lab, she held testosterone-treated females in small cages, and surprisingly, a few of these birds were able to produce wingsnaps. But an entire dance routine cannot be studied in a small cage. We have given testosterone to females and released them onto manakin leks in the wild. But unlike testosterone-treated males, who after release remain visible and active, these females slipped into the forest never to be seen, or studied, again.

Luckily, we found a solution to this problem.

There is a famous road in Panama called Pipeline Road (El Camino del Oleoducto). And when I say famous, I mean among birdwatchers. Pipeline Road is an unpaved path that runs through pristine rainforest within Soberanía National Park, midway along the Panama Canal. Extending some 21 kilometers, it was built to support the construction of a pipeline used to transfer oil between large tanker ships on either side of the narrow waterway. Since the canal was built, this forest has remained largely untouched by humans, with the exception of the road and the pipeline. At one time hunting in the forest was legal, but that is no longer the case, though poaching of larger animals remains common. Otherwise the forest is largely intact, and stunning.

Pipeline Road starts just outside the town of Gamboa, on the southern edge of the national park. When I first started visiting Panama, the road was a relatively quiet place; one could walk all day and see and hear more primates in the trees (howler monkeys and white-faced capuchins) than on the road itself—by which I mean birdwatchers, with their binoculars and cameras and spotting scopes, as well as a handful of researchers from the Smithsonian Tropical Research Institute (STRI). These biologists, mostly graduate students and postdocs, work on just about anything to do with tropical biology: they study the fauna, the flora, the soil, the atmosphere; they study things as small as microbes and as large as enormous ceiba trees. To demarcate their study areas, they put red flagging tape all over the place. But except for the tape or the occasional STRI truck tucked away on the side of the path, you would hardly know the humans were there.

Somewhat over a decade ago, things began to change. After Manuel Noriega was deposed in 1989 and a stable government put in place, efforts were made to convert Panama into an ecotourism destination. One obvious site for ecotourists was the town of Gamboa, surrounded as it was by rainforest, adjacent to the canal, and with a modicum of infrastructure in place. Two ecotourism hotels were built in or near Gamboa, and as expected—"if you build it they will come"—tourists did indeed arrive en masse. In addition, the duplex housing units in the town, built initially for canal workers, were put up for sale and the local population began to swell. Pipeline Road became very popular.

To inspire this new crop of potential nature lovers, a group of largely Panamanian environmentalists/conservationists devised a plan to construct a Discovery Center. About 3 kilometers into the forest along Pipeline Road, a path was carved in the forest that allowed vehicles and foot traffic to leave Pipeline; there the planners built several small hiking trails, a gazebo with hummingbird feeders (greedily patrolled largely by exquisite white-necked Jacobins) and posters about the local fauna and flora, and bathrooms, as well as a tower that rose some 30 meters high, providing a wonderful view of the Soberanía forest below. Last but not least, the planners built a large aviary (50 square meters in area, and 5 meters tall) to house macaws.

These large colorful parrots of the New World tropics were once abundant throughout Panama, especially the magnificent scarlet and great green macaws. Over time, however, they were extirpated from most of the interior, and today they are only found in the remote and exceedingly wild mountain forests of eastern Panama—the spectacular Darién region. Imagining that macaws could be reintroduced into central Panama, the idea was to first place them in this large aviary, and eventually open the doors and let them come and go as they pleased. Hopefully, they would depart and begin nesting on their own. Also hopefully, Panamanians living in and around the forest would be educated sufficiently that they would no longer hunt or trap the reintroduced macaws.

For some reason, however, the project was abandoned. The aviary sat unused, just 40 meters from the gazebo, surrounded by the ever-encroaching forest (see plate 12).

What if we could obtain use of the aviary and create a golden-collared manakin lek inside? We could ask all sorts of questions that were extremely difficult, if not impossible, to answer by the study of wild birds or birds held in small cages. This included, of course, whether females treated with testosterone would engage in male behavior.

The Discovery Center aviary consisted of a solid concrete base with sides of concrete, about 15 centimeters thick, that rose a meter or so above the ground. Several firm steel beams were embedded in the concrete walls, and these were surrounded by chain link fencing that formed the remainder of the cage structure. The top of the aviary, also covered with chain link fencing, was open to the air, with the exception of about 6 square meters of corrugated steel that covered one corner of the aviary, providing the occupants some protection from weather.

We pitched our proposal and were given use of the aviary for a two-to-three-year period. We were ecstatic. Because manakins could easily pass through chain link, we were allowed to modify the aviary by covering the entire structure with a durable mesh with only 1 centimeter spacing. This not only kept manakins in, but kept most other things out, like snakes, forest falcons, and wild cats.

Once the mesh was in place, we filled the concrete base with 15 centimeters of soil, and obtained small native saplings and shrubs from the STRI arboretum to plant inside the aviary. Using the proper spacing required for males to create courtship arenas, we positioned the plants in patterns we hoped would entice the males to perform their dances (see plate 13). We also equipped the aviary with four semi-ant-proof feeders (for nothing can be truly ant-proof in a rainforest) as well as several water baths.

We then released three juvenile males that had been implanted with testosterone one week previously. Our prior experience both in the wild and with small cages showed that such males would perform some courtship behaviors (wingsnaps). We added pieces of ripe fruit (manakins love papaya) and left the birds to acclimate to their new home.

The next morning we anxiously returned, hoping that they were happy. Not only did the birds seem content, but they were already displaying! Within just twenty-four hours, one male had established an arena and was performing courtship vigorously. After three days, a second male began engaging in courtship behavior, and the third male started performing rollsnaps. The three birds did interact quite frequently, with some aggressive behavior as one defended his court against the other two, and there was a bit of fighting over the papaya when it was delivered twice per day. After three weeks, this pilot program was concluded and the birds were released back where they were originally caught. Interestingly, during this time wild male manakins discovered the birds displaying inside the aviary, and two birds set up courts adjacent to the enclosure and began displaying as well. We often saw green birds (females or juvenile males) peering into the aviary, apparently watching the courtship going on inside. All in all, the aviary worked just as planned—so we now set out to do some focused experiments.

During the nonbreeding (wet) season, my colleague Ioana Chiver (an exceptional field ornithologist who found her way from Romania to Canada to Panama) captured green golden-collared manakins (females or juvenile males) and determined the birds' sex genetically. Once she had three or four birds of the same sex, she gave them implants, either filled with testosterone or blank, and then held them in small cages to acclimate them to the conditions of captivity. After one week, the birds were released into the aviary; Ioana then observed and videotaped the birds for four hours every day for three full weeks. She repeated this procedure five times, with three all-female groups and two all-male groups. In the end, she was able to observe a total of seven testosterone-treated females, four testosterone-treated males, four blank-treated females, and three blank-treated males.

The big question was, what would the females do?

The answer: not much.

To appreciate what took place, we need to consider what the males did. First, even the blank-treated juvenile males engaged in some courtship behavior: they claimed courts, did rollsnaps, and performed jump-snap displays. Those treated with testosterone did these things also, but much more so. For example, while undosed males did 2–3 rollsnaps and some 50 jump-snaps a day, those treated with testosterone upped these numbers to 10 and well over 100, respectively. Females, in contrast, engaged in almost none of these behaviors, the only exception being that one or two testosterone-treated females did produce a few individual rollsnaps. But no females, even those treated with testosterone, claimed a court, and none did the wingsnap displays of males where they jump from sapling to sapling and snap their wings in midair.

In addition to courtship behavior, we documented displays of aggression. Testosterone is known to activate territorial aggression in many animals and has been well studied in wild songbirds. As expected, some aggression, as indicated by chase behavior at feeding stations and display courts, was observed in the males generally, more so in the testosterone-treated individuals. Also as expected, blank-treated females were not at all aggressive (see plate 14). We were surprised to observe, though, that females treated with testosterone were just as aggressive as the testosterone-treated males, actively chasing one another from feeding stations and from preferred perches in the aviary. In some cases, two females grabbed each other and fell to the ground still holding on. Our docile little green females that seem to flit passively about in the dark forest understory can, with a little testosterone, be turned into bantamweight kickboxers. When this started happening, we terminated the experiment early, removed the testosterone implants, and released the birds back where they had been caught.

Finally, we examined the vocal behavior of these birds, focusing in particular on their "cheepoo" call. Here again, both dosed and undosed males produced cheepoos, and at about the same rate of 5–10 calls per 25 minutes of observation. And while untreated females never produced cheepoos, those that received testosterone did, on a general par with the males, although there was huge variation. A few females became very chatty indeed.

What can we conclude from these experiments? First, the testosterone implants had an effect on both the juvenile males and the females. Males demonstrated more courtship behavior and aggressiveness, while females became more aggressive and vocalized more.

Second, whereas all elements of the male's courtship display were activated by testosterone in males, this was not true of the females, who only did a very few snaps and rollsnaps but no other male-like courtship behaviors.

Let us now step back and consider this hormonal control of manakin behavior from the standpoint of the brain. Recall from our previous descriptions of singing behavior in songbirds as well as reproductive behaviors in other species that brains can differ across sex and age. These differences can involve the presence or absence of whole neuromuscular systems or just their neural circuits. Neural circuits can be more or less well developed in one sex or the other or in one sex across seasons. Circuits might exist in both sexes, but one sex may lack the responsiveness to hormonal activation. Hormone-responsive circuits might exist in both sexes, and the only reason a behavior is missing in one sex is that the requisite hormone is just not secreted by the gonad of that sex. How can we think about all of this with respect to golden-collared manakins?

First, it seems reasonable to conclude that both males and females possess the testosterone-sensitive neural circuitry to be aggressive. But in females it is sitting quiescent most of the time, activated only in extreme circumstances, like when another golden-collared manakin gets too close to a valuable resource. In the wild, of course, there are generally plenty of berries and perches around, so we see very little female aggression. Give the female birds testosterone, however, and those brain circuits are readily activated, such that threats to resources prompt more vigorous aggressive responses. We do not know how testosterone promotes the reactivity of this circuitry. Changing the output of neural circuits is no simple task. In extreme cases, new neurons might actually be created and incorporated into these circuits, or existing neurons might grow larger and make new synaptic contacts, connecting visual and auditory neural systems with motivational centers that also grow and make connections with the motor systems that, in this case, lead to increased aggressive behavioral output. All of these kinds of changes are examples of neural plasticity, a characteristic of the vertebrate brain with broad implications. Birds are remarkable in exhibiting considerable, even enviable, neural plasticity.

And what about the cheepoo vocalizations? Unlike aggressive behavior, females seldom if ever emit cheepoos, but they will do so after exposure to testosterone. Thus, the neuromuscular vocal systems are clearly sensitive to the hormone but likely exist in an undeveloped state in the absence of testosterone. Give the females testosterone, and the vocal circuitry and perhaps also the syringeal musculature become functional with all of the necessary anatomical and biochemical properties that require plasticity. Thus, with respect to fighting over resources and cheepooing, the key difference between adult males and females in the wild is that males have more testosterone, which naturally activates their aggressive and vocal circuits.

Let us now turn to the behavior of primary interest in our story: courtship. With the exception of a little rollsnapping, females seem completely unmotivated to perform any of the elements of the male's courtship dance—and testosterone made no difference in this. Treated or untreated, females never jumped between saplings, never did a single midair wingsnap, never did a grunt-jump display. They also never tried to mount and copulate with other females. Thus, we might conclude that females completely lack the circuitry that enables male courtship or copulatory behavior. It is also possible that females have the circuitry but that the enzymes or receptors that would allow the circuit to respond to testosterone are lacking. Because the bulk of these behaviors seem never to be performed, it is likely that the circuits disappeared in the developing female brain (i.e. were demasculinized) or were generated to grow only in males (i.e. were masculinized).

While this perspective seems reasonable, there is an important caveat. Bear in mind that once a choosy female finds a male to her liking, she joins him in the display, descending to his court and flying (not jumping) between saplings in much the same way as does a male. She does not make midair wingsnaps, she does not do a grunt-jump display, but she is clearly physically able to move herself between saplings. Females have at least some of the necessary neural and muscular systems to perform elements of the dance, but rather than going solo, a female uses her physicality to join the male in a pas de deux. What is missing in the female, it seems, is the motivation to perform in the absence of the male's display. In other words, she lacks the steroid-sensitive neural motivation system to perform on her own.

When present, where are these circuits in the brain? We suspect that the premotor circuitry that drives male courtship lies in some of the same brain regions that control singing behavior, in the subdivision of the forebrain called the telencephalon. Areas homologous to our own motor cortex reside in the avian brain and include areas we have described for song. One such region, the arcopallium, the region where the oscine song nucleus RA is found, has also been shown to control some wing movements. In male manakins, the area is rich with androgen receptors. Hence, hormone-sensitive neural circuits likely exist in the arcopallium to drive the neuromuscular systems of the magnificent courtship display.

But what about the motivation to perform courtship? We need to introduce a new, but extremely important brain area: the hypothalamus.

The hypothalamus is an extraordinary structure. It arose early in vertebrate development, becoming the site where much of the body's physiology is controlled, including reproductive physiology. As described earlier, if the anatomical and physiological systems are a go for reproduction, then you want reproductive behaviors to be activated as well. That occurs in the hypothalamus. It contains the neural circuitry that, when activated, leads to the motivation to reproduce. This region also contains some of the circuitry that gives rise to aggressive behavior. Moreover, whereas we know that the neural system controlling song learning and expression is complex, is associated with auditory centers, and is sensitive to sex steroid hormones, apparently the motivation to sing lies, at least in part, within the hypothalamus. If this region of the brain is singled out for treatment with sex hormones, birds sing more, but their songs are lacking in their full richness.

Collectively, this work offers the perspective that, like song, motor systems underlying male manakin courtship are found in their large, expansive telencephalon, whereas the motivation to perform courtship rests in circuits found in the male's hypothalamus. Our data suggest further that female manakins possess a hypothalamic hormone-sensitive center that motivates aggressive behavior, but they lack the hypothalamic circuitry for courtship.

This chapter has focused largely on courtship displays proper. But the male display is performed around an arena, a piece of land that the bird claims for himself. Sometimes he claims two to three such courts, all within sight of one another. The bare dirt in the middle of the court is always clean, in the middle of rainforest! What is a court? What does the male see, and how does he maintain it? This is the subject of the next chapter.

TEN

Male Manakins Keep
Their Gardens Clean

Despite the raucous sounds and brilliant colors of male golden-collared manakins, they can be quite difficult to see in the dense undergrowth of their forest homes, especially at midday, when they are decidedly still and quiet. The best strategy is to focus on the forest floor. What you are looking for is a bare patch of dirt, approximately "two and a half feet long by twenty inches in width," as Frank Chapman described a manakin arena in 1935 (see plate 5). Why is such a spot notable? Consider the typical rainforest, its floor covered in tree litter—dead leaves, twigs and branches, flower petals, fruits, etc. If a patch of dirt becomes exposed, say when a tree falls over, the bare dirt quickly sprouts back to life with lichens, mosses, small herbaceous plants, or saplings of great trees. Bare dirt is a rare, but important, commodity. And sitting somewhere near a clear patch of earth in an otherwise undisturbed forest will be the gardener who keeps it clean: a male golden-collared manakin.

Somehow, these little birds defy the forces of the forest that want to fill in that gap. How do they do it? As Chapman observed, "The bird removes leaves and other material from its court with its bill. This act seems to be more or less sexual in character and often terminates the court display of snapping. Small leaves are carried to a height of about three feet and dropped at an equal distance from the court. Large ones are taken in a more direct line to the border of the court. One that I saw removed measured 10½ × 3⅜ inches; it was therefore slightly more than two and a half times as long as the bird that carried it."[1] We have also observed that males can keep the space directly above the court clear by continually picking at any leaves that might try to grow there.

Males are highly motivated to keep their court very clean, and testosterone makes them even more fastidious. Females, in contrast, seem completely uninterested in gardening, even if they are given testosterone.

We know this from the forest aviary. Here, we planted saplings to simulate the natural forests where males establish courts. As mentioned previously, we trucked in soil to fill the concrete base about 15 centimeters deep throughout the entire 50 square meters. As it turned out, either the soil itself contained some seeds and spores, or they blew into the aviary and took root in the fresh soil. Most were tiny. The males, however, did not like them at all—at least those that germinated between the small saplings that made for a court—and plucked at them repeatedly. Eventually, without their little leaflets, the sprouts disappeared, and the court was again perfectly clean.

The aviary's roof was made of chain link fencing and a fairly fine plastic cloth mesh; as a result, leaves could not reach the floor, even from an imposing fig tree overhanging part of the aviary. So we added some. Every day, Ioana Chiver collected dried leaves and scattered them about the aviary. Any leaves that she dropped onto the courts were quickly picked up and carried away. Juvenile males without testosterone did some of this tidying up, but more leaves were removed by males with testosterone. One male created a leaf pile about half a meter high, a good meter from the center of the court. No female was ever seen to pick up a leaf or perform any other gardening behaviors, even when treated with testosterone.

Ioana examined this leaf-clearing behavior in wild manakins too, who do not have the benefit of a protective roof. The diversity of leaf sizes and shapes that grow on trees and shrubs in the rainforest is quite remarkable, the result of adaptations that allow moisture, nutrients, and sunlight to be gathered and protect the plant from insects and other herbivores. Some leaves are huge, as large as the area encompassed by a person's arms held out in a circle. Others can be tiny, only a few centimeters long.

Most rainforest trees are at least somewhat deciduous—that is, they drop and replace leaves on a regular basis. Generally speaking, these trees thrive during the rainy season, when it may rain many centimeters a day. The dry season, in contrast, can bring weeks with no rain. In such times of drought, the trees drop some or all of their leaves. The dry season is also the time when male manakins court females. So the same testosterone that tells them to perform their courtship displays also tells them to be enthusiastic about cleaning up all those falling leaves.

Just how big a leaf can a manakin move, and how does he do it?

Ioana tested this question by buying an artificial potted plant, removing some of its identically shaped fake green leaves, and placing them in the middle of the courts of a number of wild male manakins. Only there was a twist: on the underside of each leaf she taped a small weight. A single leaf with no weight attached weighed approximately 1 gram. (By way of comparison, a U.S. dime weighs a little over 2 grams.) At first, Ioana taped thick metal wire to the leaves, but she soon realized she needed heavier weights, so she turned to, appropriately enough, coins: pennies (weighing 2.5 grams), nickels (5 grams), and quarters (5.67 grams). Recall that the average male manakin weighs approximately 18 grams, the equivalent of three quarters. Having set up a video camera, Ioana placed a leaf on a court. She returned four hours later to see what happened.

For the most part, a male responded to the leaf within minutes after Ioana departed the arena. Indeed, it was virtually the complete focus of the bird's attention. Sometimes he would jeer at the leaf and make wingsnaps. He seemed angry it was there. Then he would perch (sideways) on the sapling closest to the leaf, leap off, grab the edge of the leaf in his beak, and fly outside the court, where he released the leaf—assuming it was light enough to be lifted. Males could easily pick up and fly off with a leaf weighing between 1 and 10 grams, but they also had pretty good success with leaves weighing 12–16 grams, and occasionally even 18–22 grams.

As leaves got heavier, they posed a problem, but one that could usually be solved. If the bird could lift the leaf but not actually fly off with it, he would simply lift it and move it a short distance, over and over until he reached the edge of the court. Heavier leaves, 24 grams or more, could not even be lifted, so they were dragged. The heaviest leaf receiving this treatment weighed 36 grams. A leaf of 50 grams could not be moved at all. Because we worried a male would abandon his court if presented with a completely recalcitrant leaf, we made very few tests of these heaviest leaves. The fact that at least one male could move a leaf twice his body weight was compelling enough data for us.

There are several considerations in all this that I find fascinating. What is clear is that only males claim courts; females do not. Moreover, males keep their courts clean, while females contribute none of the gardening behaviors that males so readily display. Males are incredibly attentive to the bare dirt of the display court. When, in our experiment, a leaf was placed on a court, the male's attention shifted to the leaf. While some males were nervous when we were present, others completely ignored us and other stimuli in the environment, remaining totally focused on the leaf intruder until it was removed. We captured one video of an agouti, a large forest rodent, walking by an arena in which a weighted leaf had been placed. The male manakin remained vigilant on a nearby sapling, eyeing the leaf, eyeing the agouti, the leaf, the agouti, until the mammal had passed by—at which point the male dropped to the dirt, picked up the leaf, and flew off with it.

This description of gardening is reminiscent of what we have described previously for several sex-specific behaviors—behaviors activated by testosterone. Discrete neural circuits, which are often sensitive to sex steroids, control many sex-specific behaviors associated with reproduction. By extension, we assume that specific neural circuits must control gardening behavior. But where do they reside? Are they in sensory systems? Are there special visual processing centers that recognize bare space on and above an arena? If that image is disturbed, do those visual processing centers project to and activate motor systems to promote removal functions? Are these sensory and motor centers activated by testosterone, or are there discrete gardening motivational centers located in the hypothalami of male manakins? Does the female have the same visual processing system that sees the bare spaces of the arena, and does she use this information, in part, to assess the quality of the male, but without the circuits being activated to promote leaf removal? In other words, does she simply lack a leaf removal motivation center while possessing a quality judgment center? Does such a quality judgment center activate the motor patterns to perform improvement behaviors in males, while in females it activates sexual motivation centers?

Some of you may be yawning and saying, "So what if manakins move some leaves around? What about bowerbirds of Australasia and those males' architectural masterpieces?" I agree, bowerbirds are extraordinary. Nevertheless, the questions we are asking are virtually the same for manakins and bowerbirds—and, by extension, all sorts of animals, probably us humans as well. To wit, where are the neural circuits underlying the motivation and capacity to perform sex-specific gardening or arena construction behaviors? Where in the CNS are the neurons that are activated by testosterone, and just how much of that circuitry is shared by males and females?

What makes all of these behaviors and their associated neural circuitry so fascinating is that they seem to reflect a relatively complex form of cognitive function that mediates sensory input and motor output. A chick might see a red spot on a parent's beak and peck at the spot to be fed: a sensory perception translates to a motor reaction. In the case of the manakins, the visual stimulus is three-dimensional space, which requires regular assessment and appropriate action.

But what if the offending leaf is heavy and the male manakin's initial appropriate action doesn't work? Fortunately, he is adaptable enough to try a different strategy: if he cannot fly the leaf out of his arena, he'll stay on the ground and hop it out. It is hard not to anthropomorphize, but the male seems to pause and consider the problem in order to devise the next approach. Of course, when we humans "consider" a problem to be solved, we might actually picture what motor actions are needed, which we then perform in hopes of completing the task. We do not know what is going on in the bird's brain; we only know that they perform an initial motor action, and if that proves insufficient, they pause and perform another motor task. Call it what you will, but I call it a significant cognitive function that has profound sophistication.

Birds solve problems. We know that ravens and crows and some jays solve tasks posed by scientists to assess their intelligence or cognitive capabilities. We know that autumn nutcrackers can collect thousands upon thousands of conifer seeds or nuts, store them in hundreds of locations over enormous areas of forest or mountaintop, and then remember those locations, even when they become covered with snow. We consider this group of birds, the corvids, special because of these behaviors. I think manakins are special also. In fact, manakins aren't all that different from other birds: they go about their day solving tasks, many of which are quite complex, until they finally rest at night. Since berries, the preferred food of golden-collared manakins, are usually readily available, they don't need to focus on finding and storing food; instead they have plenty of time during the day to perform court maintenance. They deserve our respect for this impressive cognitive skill.

The leaves we used for testing were both long and wide, 12 centi-meters by 8—in other words, as long as the bird and wider than he was wide (unless his wings were fully spread). In addition, the leaves weighed up to 60 percent of the bird's own weight, and sometimes as much as 90 or even 120 percent. And yet they were still able to fly carrying this unwieldy, heavy leaf—and do so using only their beak. By way of comparison, imagine a 68-kilogram human picking up—by their teeth!—a 2-by-1.5-meter piece of plywood that (impossibly) weighs 40–82 kilograms and jumping into the air with it. Perhaps some Olympic weightlifter on steroids could do this, but they'd bet-ter have a good dentist and chiropractor.

The fact that a male manakin picks up an object as much as or more than his own body weight, in his beak, and then flies should seem like a good enough trick. As we have seen, one adaptation enabling effective flight involves the accumulation of weight at the bird's core, thus placing its center of gravity between and beneath the wings. Weighted structures at the bird's periphery are also min-imized. Thus, birds' tails, wingtips, and legs are all lightweight, while the muscles that control them are enlarged centrally. This is true also for the head: birds have eliminated the need for heavy teeth by means of a muscular stomach, or gizzard, in their upper abdomen, which grinds their food for digestion; instead of teeth, then, they have a lightweight beak, which also allows the neck and head muscles to be somewhat reduced. But that still doesn't explain how a male manakin can carry heavy leaves.

Let's look at some other birds to explore this issue. Many shorebirds that breed in the arctic, or warblers that breed in North America, in fact double their body weight with added fat prior to performing long-distance autumnal migrations. And hawks, eagles, ospreys, and owls easily pick up large prey items and fly off with them. These birds seem quite capable of carrying heavy weights. Are manakins really all that special?

In the case of the shorebirds and warblers, where is the fat actually stored? As you might suspect, it's in and around their core—between their furcula, or wish bone, in their thorax, or on their flanks and upper abdomen, mostly beneath and between their wings. In the case of shorebirds, long wings adapted to life in open country assist their flight. The warbler, though, like the manakin, lives in forests and is designed for mostly short bursts of flight. But when it's time to breed, it fattens up on berries and then carries this weight for 4,000 kilometers. Again, for efficiency, this mass is stored beneath and between the wings. (Come to think of it, compared to the manakin, the warbler's feat might be a tad more impressive.)

What about all those raptors that fly off with their prey to consume it in a protected tree, cliff, or nest? These birds, too, are designed for flight, with long broad wings to keep them aloft on rising columns of air. And their imposing beaks are in fact fairly light structures, used not for killing but for tearing at a carcass. Instead, their prey are dispatched by a powerful squeezing of the talons, which crushes the hapless victim. Their real weapons are their legs, and the massive muscles that power the talons. All of the weight needed to kill and carry prey is located, you guessed it, between and beneath the wings.

Have you ever seen an osprey capture a fish and carry it off? As big as these birds are (1–2 kilograms on average), it's not an easy task.

Although the fish an osprey catches are usually somewhat smaller than it is, 0.5–1 kilogram (or exceptionally, 2 kilograms), still the bird struggles to take off from the water and gain sufficient altitude to reach a nearby perch to consume its prey. To help in this effort, it carries the fish head first (not sideways), which improves the aerodynamics and efficiency for flight.

The great horned owl of the Americas is considered a fearsome predator of a huge variety of prey. Whereas the owls weigh, on average, about 1.5 kilograms (females are larger than males), they regularly eat prey—often some sort of rabbit—weighing up to about 2.5 kilograms (though they have reportedly taken prey up to 6 kilograms in weight). The most powerful raptor in the world is the harpy eagle of Central and South America. Weighing 9 kilograms or more (again, females are larger), they regularly attack, kill, and carry prey such as monkeys and sloths that can nearly equal their mass. Consider a harpy eagle carrying a 6-kilogram spider monkey. Now go over to your set of free weights and pick up 15 kilos. Heavy, right? Think of these eagles taking to the air with all that weight. Amazing! But again, the prey is carried in the talons, which lie beneath the center of the bird, keeping the center of gravity beneath and between the wings. The same is true of the huge muscles that power the lifting and depression of those wings. The design is altogether perfect. Unless you're the monkey.

These examples illustrate the point that when birds carry weight, they do so in a manner that maintains efficiency for flight. Male manakins may be small, but they, too, pick up relatively heavy and awkward leaves with their beaks and fly with them. Keeping their courts clean so females can watch them dance requires strength. Did their musculature evolve in response to this need (and others), or did these birds evolve various uses, including gardening, for their innately powerful muscles? As we will see next, steroid hormones play a crucial role in establishing the capabilities of at least some of the male manakin muscles.

ELEVEN

Male Manakins Are Rich with Androgen Targets

The complex essence of male manakin social and sexual behavior is, as we have seen, largely activated by androgens. To see just how these powerful signaling molecules exert so many effects on the male's life, we need to take a deeper dive into the ways that hormones capture an animal's organ systems, most importantly the brain and the muscles it controls.

When we matter-of-factly say that a hormone acts on some target tissue, as when testosterone acts on the brain or a muscle, what do we mean? It's not magic. In actuality, it is an important function that gets directly at the whole concept of what it is to be an animal cell (or, for that matter, any cell of a multicellular organism), with DNA and associated processes instructing that cell exactly how to behave. As a reminder, with few exceptions, each cell in our body houses chromosomes composed of long chains of DNA, the genetic code that instructs all life events. Some of those chains of DNA are compartmentally organized into genes, with each gene providing the instructions for making one or more specific proteins. The instructions are first read and encoded in messenger RNAs, which are converted into proteins by a process called translation. It is the proteins that do most of the work in the cell, and the type and number of the proteins that are present is what makes each cell unique. All of your trillions of cells are doing this all the time, as are all of the cells in a manakin.

The extent to which any gene undergoes transcription—that is, whether it undergoes transcription at all, and if so, by how much—is determined by molecules called transcription factors. While every cell in the body contains essentially the exact same DNA, cells vary profoundly in their structure and function by virtue of the transcription factors present in them, which determine the extent of transcription of all the genes in that cell. A neuron, with its vast array of slender dendrites and lengthy axons and which communicates by fluctuating electrical potentials across its membranes and the release of neurochemicals from its axon terminals, has the same DNA as a skin cell, or liver cell, or cell in the kidney or pancreas or colon or big toe, cells that may never change their membrane potentials and may never secrete a thing. Control of transcription is thus key to the evolution of multicellular organisms possessing unique cell types, as well as to what makes us us, and manakins manakins.

Thus, any mechanism that gains some control over the transcriptional process possesses great power. This is exactly what steroids do. Upon binding to its receptor, a steroid hormone converts that receptor into a powerfully active transcription factor that, by controlling the expression of various genes, gains control over the functions of many cells. This is what testosterone does; but just how the hormone testosterone influences behavior will be a surprise to many.

In certain respects, testosterone is not a hormone at all, but a prohormone—that is, a precursor to one or more other hormones. As discussed previously, enzymes (proteins that catalyze biochemical reactions) start with cholesterol and synthesize steroid hormones,

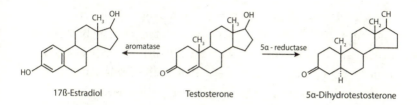

Fig. 11.1. Two-dimensional structures of testosterone and its active metabolites 5α-dihydrotestosterone (DHT) and 17β-estradiol. Drawing by Bill Nelson.

with some of these steroids serving as substrates for the enzyme-mediated synthesis of products that are different kinds of steroids. By these reactions, cholesterol is converted into pregnenolone, pregnenolone can be converted into progesterone, and progesterone can be converted (by two enzymes) into testosterone. Thus, one can see that progesterone, a molecule that we have already identified as an active steroidal hormone that binds to its very own set of receptors, can also serve as the substrate for the formation of a completely different set of steroidal hormones, namely the androgens, which include testosterone. Were testosterone to act fully on its own, we'd call it a hormone and the story would be over. Yet the cascade of reactions does not stop there. Indeed, two enzymes can act on testosterone to create vastly different hormonal signals (fig. 11.1). One enzyme, 5α-reductase, converts testosterone into 5α-dihydrotestosterone, or DHT. DHT is a potent hormone that binds to androgen receptors more potently than does testosterone. In many male tissues, DHT is *the* androgen that promotes all things masculine.

Then there is the fact that testosterone, usually considered the principal hormone of males, upon interacting with an enzyme called aromatase, is converted into estradiol, the principal hormone that we associate with females and femininity. This reaction occurs in the female ovary, which both makes a good amount of testosterone and also expresses a considerable amount of aromatase. The resulting estradiol then gains access to the bloodstream and circulates at elevated levels during discrete periods of female reproduction. In the chicken, as we have seen, estradiol is secreted into the bloodstream around the time of ovulation, where it acts on oviduct, liver, and bone to make an egg. By at least one measure, all of this starts with the production by the ovary of the androgen prohormone testosterone.

It turns out that many male tissues and organs also have aromatase, as well as 5α-reductase. When circulating testosterone reaches these tissues, males can make either the potent estrogen estradiol or the potent androgen DHT. Indeed, males of essentially all vertebrates, from fish to mammals, including us, have in some parts or other of their bodies the enzyme 5α-reductase that makes DHT as well as the enzyme aromatase that makes estradiol. Moreover, not only do male tissues have the receptors for androgens, but some also have receptors for estrogens. Again, virtually all vertebrates, from fish to mammals, including us, have, in some parts or other of their bodies, androgen receptors and/or estrogen receptors. This leaves us with a surprising situation: when considering the action of testosterone in males, we need to ask, is it acting as an androgen or as an estrogen?

Although our focus here is on males, they are not alone in these endocrine features: some female tissues and organs also have the enzymes that make potent androgens and estrogens, as well as the receptors for both androgens and estrogens. Thus, the widely held perspective that rigidly couples androgens with maleness and estrogens with femaleness is strikingly oversimplified.

With this in mind, how, then, does testosterone activate the complex golden-collared manakin courtship behavior? Put another way, does testosterone activate male behavior by first being converted to DHT for action on androgen receptors, or is testosterone converted by aromatase to estradiol to act on estrogen receptors? Perhaps both androgen and estrogen receptors require activation. This question is not just simple curiosity. In many species, the male's brain has aromatase and estrogen receptors, and behaviors like aggression and copulation are activated not by androgens but by estrogens. The manakin brain has both aromatase and 5α-reductase, so to understand hormonal control of manakin behavior we need to know which steroidal pathway is involved.

Knowledge of the receptor mechanisms for steroidal action gives us a way to find out. Here, pharmacology, often driven by a hungry pharmaceutical industry eager to find cures for cancer, is of help. It turns out steroids, in particular the sex steroids like androgens and estrogens, are very potent at triggering events that produce cancers. In women, mammary tissue and uterine tissue are especially prone to developing malignancies due to estrogen actions on cells that are out of whack. Similarly in males, prostate cancer often arises because androgen signaling goes awry. One way to attack these steroid-dependent pathologies is to block the action of each steroid type on its particular receptor. That is, if we had a molecule (a drug) that could prevent androgen from binding to an androgen receptor or estrogen from binding to an estrogen receptor, we could inhibit the growth or spread of androgen- or estrogen-dependent tumors. These kinds of drugs are called antagonists or inhibitors, and they can be used for more than inhibiting disease. They can also be used to identify the natural actions of specific hormones on target tissues or on behavior.

We suspected that testosterone in manakins largely acted through androgen receptors. We were able to test this by treating adult breeding male manakins (that is, males performing courtship on their own courts inside their own leks) with an androgen receptor antagonist. My colleague Leo Fusani and I chose to use a compound called flutamide, which we could give to the manakins by implants similar to those used to give testosterone. After receiving an implant, the birds were released back onto their leks and their behavior was observed daily for about three weeks. What we saw first confirmed our suspicion that androgen receptors were involved in their courtship behaviors. What we saw second was altogether different, however, and suggested that our interpretations might be overly simple. In the first week, birds implanted with flutamide displayed significantly fewer jump-snap displays than control-treated breeding males, as well as fewer wing- and rollsnaps. A week later, the difference had disappeared, as if the flutamide was no longer working to block androgen actions. A week after that, all the treated birds performed significantly *more* jump-snap displays and wing- and rollsnaps, as if flutamide was now acting as an agonist on the androgen receptors— that is, not blocking androgen action, but actually stimulating the receptors. These data were hard to interpret, but one way or the other we could conclude that androgen receptors were indeed playing a role in manakin courtship.

So, where are androgen receptors found in manakins?

We might well assume that androgens act on the brain to stimulate males to perform courtship displays. While this may be true, is the brain the only place that androgens act? Recall that there are really two (or more) ways in which hormones might stimulate behavior. The first increases the *motivation* to perform the behavior, while the second increases the *capability* to perform it. Motivational centers are found in the brain, with the hypothalamus being one site where androgens might activate neural circuits that form the basis of motivation. The actual capacity to perform behavior, however, resides in many places. The motivational centers thus need to connect with those neural circuits that actually create the muscle contractions that constitute behavior. There are premotor neural circuits in the brain that require activation and these might be targets of androgen action. Premotor neurons relay information to motor neurons mostly located in the spinal cord. Thus, androgen might act directly on spinal motor neurons. These neurons project to and stimulate contractions of specific muscles that generate the movements of a specific behavior. Thus, these skeletal muscles might also be targets of androgen action. This leaves us with the prediction that for male manakins to perform their elaborate courtship behaviors, we would expect to find androgen receptors in the hypothalamus, in nonhypo-thalamic premotor brain regions, in the spinal cord, and in those skeletal muscles underlying all of the movements we witness as courtship.

Our data show that, indeed, manakins have androgen receptors in each of these regions.

There are several ways to determine if a tissue or cell has receptors for a hormone. One can take the tissue, homogenize it in a buffer, add some radioactive hormone, incubate it for a while to let the radioactive hormone bind to the receptor (if any is there), then separate the bound radioactive hormone from the free radioactive hormone. If any radioactivity is in the "bound" fraction, then there is some kind of substance, presumably a molecule, holding on to that radioactivity: a receptor. Although this method may tell you whether a receptor exists, it does not provide much detail about which cells have the receptor or where they might be located within the tissue that was homogenized.

Another approach applies the same concept—that is, binding of radioactive hormone to receptor—but uses a mechanism that gives some idea of the anatomy of the tissue. In this case, one takes tissue that has been frozen or otherwise "fixed" with a chemical to preserve the tissue structure, and then cuts a very thin slice. Some radioactive hormone in solution is placed on the slice, which is again incubated to let the hormone bind to any receptor molecules; the sample is then washed to remove unbound radioactive hormone. The procedure now moves into a darkroom, for the next step involves placing the thin tissue slice on a glass slide that has previously been dipped in photographic emulsion. And there the slide sits for weeks, even months, until, still in the dark, the slide is "developed" just like photographic film. Because some radioactive decay emits beta particles that resemble photons, they act like a light source to expose the silver grains in the emulsion. Those silver grains lying directly beneath cells on the tissue where radioactive hormone has bonded with receptors will become darker over time. When the sample is viewed under a microscope, the exact position of those cells can be visualized. This procedure is called autoradiography.

We used this technique in our initial study of golden-collared manakin androgen receptors. Our first question focused on the spinal cord. We anticipated that because so many skeletal muscles come into play in the male manakin display, and most of those muscles are innervated by motoneurons in the spinal cord, the cord might be especially rich in androgen receptors in males, and perhaps less so in females.

So I traveled to Panama with my friend and ornithology colleague Fritz Hertel (doppelganger of the "The Dude" in the Coen Brothers' film *The Big Lebowski*), and we set about capturing adult manakins of both sexes. This was my first attempt not only to capture and hold wild birds for experimental purposes, but also to collect tissues in Panama and return them to my lab at UCLA for processing and analysis. Each step proved challenging—and at times almost humorously catastrophic.

Once birds were captured, we injected a prescribed amount of radioactive testosterone into each bird. We used testosterone labeled with tritium, a radioactive form of hydrogen that, upon decay, emits very weak beta particles that are very safe, traveling in air for only a few millimeters and which are blocked by just the thin outer layer of skin. Bear in mind that if receptors are bound by endogenous hormone, they will be unavailable for binding to the injected radioactive hormone. So we needed to remove the birds' endogenous androgens, presumably produced by their testes. As I mentioned previously, it is difficult and stressful to surgically remove the gonads of wild birds, so we did the next best thing, which involved injecting the birds with a drug to block androgen synthesis. (The drug we used, ketoconazole, may be familiar to some of you. If you've ever had a topical fungus like athlete's foot or jock itch, it's possible you used an antifungal cream. And the active ingredients in such creams all end in -*azole*; you might even find ketoconazole itself on the label. For some reason, this class of drugs not only inhibit fungal growth, but they also block the activity of 17α-hydroxylase, the enzyme required for androgen synthesis. Go figure.)

Once androgen synthesis was blocked, we injected the birds with radioactive testosterone, waited an hour or so, then sacrificed them, removed their spinal cords, and stored them in lead-lined containers in an ultrafreezer (–70° C). When all the birds had been processed, we packed everything up, including plenty of dry ice for transport. We would fly to Los Angeles, with a five-hour layover in Miami; we would then proceed directly to UCLA, where the tissue samples would be transferred to another ultrafreezer for safe storage. So far, so good.

But then we hit U.S. customs in Miami.

When I told the agents we had wild bird tissues, they immediately whisked Fritz and me, and our large cardboard box filled with extra bird cages and a dry ice–filled cooler containing tissue samples, to a small room. When asked if I had a permit, I explained that I had been careful to check on this, and the U.S. Fish and Wildlife Service had told me none was necessary. I was informed that I did indeed need a permit—not from Fish and Wildlife, but from the U.S. Department of Agriculture. Why did the USDA care about a wild bird? I asked. Panama, it turned out, had reported cases of Newcastle's disease, a serious problem for domestic chickens. And since wild birds can be a vector for Newcastle's, we could not bring the tissue in without a permit from the USDA. Our samples would have to be impounded.

I explained that if the dry ice evaporated, the samples and all our work would be ruined.

They didn't care.

I pointed out that we only had frozen samples from eight birds, but hundreds of millions of birds migrate between Panama and the United States twice each and every year. To which the customs officer replied: "Well, we can't stop them. But we can stop you."

Hmmmm. I decided to try a different tactic. I asked if there was any way I could get the permit then and there.

They scratched their heads and said, "Well, maybe." The problem was, it was Sunday.

But they let me use their phone, and somehow I was able to make contact with a UCLA lawyer. I pleaded for help, and he said he'd try. We had about three hours before our flight to Los Angeles.

After an hour of calls between the UCLA attorney and the customs officials, who, starting to feel sorry for us, had decided to help, it appeared a deal could be worked out. We could obtain the permit, which cost $50, and pay a $25 fine, and they would let us go. However, they only took cash, money order, or bank transfers, *and* the payment had to go to Washington, DC. Aaargh! It was now just one hour until our flight.

Many more desperate phone calls later, the USDA finally agreed to accept a credit card. All the card details had to be given twice, once for the permit, once for the fine. With fifteen minutes before departure, the customs officials had a cart waiting to take us to the gate and our box of samples to the plane. Upon dropping us off, they bid us goodbye with a smile. We were all friends by that point.

Fritz and I boarded the completely full 747 and took our seats near the rear. Looking out the window, I saw a tractor pulling a cart; on it was one item: our cardboard box. I breathed a sigh of relief. The baggage handler picked up the box to put it on the conveyer belt, but then stopped, set the box on the ground, and peered inside. He called over the driver of the tractor, and they both peered inside, talking. Then they put the box back on the cart and the tractor drove off, back under the terminal. "Oh no!" I yelled. In a panic, Fritz and I ran the full distance of the plane, just as the doors were about to close.

This was in 1997, before the age of terrorism. There is no way we could have gotten away with this now. Today, I am sure, we would have been tackled, duct-taped, and arrested. But not back then. We begged the flight attendant to listen to our story. Eventually she agreed to make a call to find out what had happened to the box. We had to return to our seats, under the suspicious scrutiny of the hundreds of other passengers onboard.

After a few minutes, the flight attendant came back to inform us that, upon looking in the box and seeing bird cages, the handler thought it would be better positioned in a compartment at the front of the aircraft where fragile items were kept. The box was safe and sound, she said, and definitely on the plane.

The plane took off only a few minutes late. I am quite sure I had several of those little Dewar's bottles on that flight.

Upon arrival in Los Angeles we headed straight to UCLA, where the samples could be returned to an ultrafreezer. Our lengthy adventure was finally over. A few days later, we sectioned the spinal cord tissues and, in the dark, placed them on emulsion-coated slides. We then stored them in light-tight ammo boxes under refrigeration for six months. Finally it was time to develop the slides. We dipped them in a staining dye and looked at the slides through a microscope. Sure enough, there they were: small, tight clumps of silver grains located directly beneath relatively large stained cells. The experiment had worked! Receptors, presumably androgen receptors, were present in motoneurons in the manakin spinal cord, holding tightly to the radioactive testosterone that we had injected in Panama.

We didn't just smile and go home when we saw these androgen-sensitive cells. No, I enlisted a new graduate student in my lab, Doug Schultz, to count all of the cells in all of the spinal cord slices. In the end, he found that there were many more silver grain clusters in the spinal cords of male manakins than in female manakins. Indeed, females had very few of these cells at all. Could this be one reason why males naturally perform male courtship displays, but females do not—because females lack androgen-sensitive motoneurons in their spinal cords? The simple answer is, probably not. A crucial question remained: just what muscles do these androgen-sensitive motoneurons control? Do they have anything to do with courtship behavior?

To tackle these latter questions, we needed to employ an entirely new set of procedures. Recall that proteins are produced in cells when DNA undergoes transcription to make messenger RNA, a specific nucleotide sequence that is then translated into an amino acid sequence that is a protein. Steroid receptors are proteins, so cells that possess androgen receptors must first make androgen receptor mRNA. One can obtain an indirect measure of the amount of androgen receptor in a cell by documenting the amount of androgen receptor mRNA present. Recall, too, that all cells have roughly the same DNA and so all cells have the sequence encoding the androgen receptor. However, each cell has its own unique capacity not only to transcribe that DNA into mRNA, but also to determine by how much. Thus, some cells may have a lot of mRNA for androgen receptors, some may have a little, and some may have none. Some cells may have mRNA for androgen receptors at only some times of the year and not at others.

One way to assess the amount of mRNA expressed in a given cell is by a procedure called in situ hybridization, where another nucleotide sequence, specific for some of the androgen receptor sequence, is made that not only will hybridize to the androgen receptor mRNA in the cell, but can also be visualized by some method, including by creating that nucleotide sequence with a few radioactive nucleotides. Once hybridization occurs, the process becomes something like the autoradiography described above for hormone binding. Any cells with radioactivity will expose silver grains on an emulsion-coated slide, and we will know those cells possess androgen receptor mRNA.

The autoradiography involving hormone binding required us to stain the tissue to see the cell bodies that sat on top of clusters of silver grains. But one can use another trick. Rather than stain those cells after the tissue is on the slides, one can fill the cells with dye, which can then be visualized by applying specific kinds of light.

This is what we did with some of our manakins. Prior to sacrificing the birds and removing the spinal cords, we injected one of two special dyes into specific muscles of interest. Because we were interested in identifying the motoneurons that innervated the major forelimb muscles involved in manakin wingsnapping, the SH, SC, and PEC muscles, we injected those muscles with either "fast blue" or "fluorogold" dyes, which are known to be picked up by the nerve terminals in the muscles and transported backward (retrogradely) up the axons to the somas (bodies) of the motoneurons located in the bird's spinal cord. It is these cell bodies that might also house any mRNA for androgen receptors. Our experiment, therefore, involved injecting dyes into the muscles of interest, waiting three days for transport of the dye to the spinal cord, then sacrificing the birds and removing the spinal cords; these were then sectioned and exposed to a radioactive nucleotide sequence that would hybridize to androgen receptor mRNA; after allowing the radioactivity to expose silver grains, we developed the tissue slides; then, using a fluorescent light, we examined the slides to see if clusters of silver grains were found underneath individual cells that fluoresced either blue or gold (which just happened to be UCLA's colors!). This would tell us if manakin wing muscles are innervated by motoneurons that express androgen receptors.

This is precisely what we observed, but even more. Not only did the motoneurons fluoresce and express androgen receptors, but so did many cells in dorsal root ganglia. Ganglia are clusters of neurons that lie outside of the central nervous system. The dorsal root ganglia lie alongside the spinal cord; they house the somas of the sensory neurons that receive stimulation from peripheral sensors and transmit information to the spinal cord. These instructions about the position or contraction state of the skeletal musculature can regulate the activity of motoneurons while also projecting upward to the brain to inform the entire central nervous system about the positioning of all muscles in the body.

This sensory communication is crucial for behavior. In order to move a limb to the proper position, a muscle must know exactly where that limb is in space, and it must contract with extraordinary precision. The sensory information being received is of profound importance. Our appreciation of the continual need for feedback information of a muscle's contraction state grows as the degree of movement expands across time and becomes more complex. Consider the control needed when two muscles are contracting, or contracting and relaxing simultaneously, to create an intricate limb movement. Each of those muscles must know exactly where that limb is in space and across time based on the contraction state of the other independent muscle. When walking, the left leg and its muscles need to know what the right leg and its muscles are doing, and vice versa. Now imagine all the muscles that are contracting or relaxing as a manakin performs its elaborate courtship display, as limbs and torso, neck and head, all change positions at great speed. A marvelous collection of neural motor and sensory neurons are at play.

What we found in the manakins is that not only did the motoneurons innervating the forelimb muscles express androgen receptors, but so did many, if not all, of those sensory neurons. We later discovered that the spinal cord was rich with the enzyme 5α-reductase, so, in all likelihood, the peripheral neural control mechanisms underlying male manakin wing movements were under some control by circulating testosterone being converted to DHT for action on androgen receptors. As these neurons all control appropriate motor output, they reflect a role for androgens in the birds' capacity to perform behaviors that involve the wings, movements like wing- and rollsnapping.

But how does the motivation to perform courtship enter into all this?

■ ■ ■

No doubt, the activated neurons that stimulate any motivated behavior are located in the brain, probably the hypothalamus. As described previously, we know that the hypothalamus contains nuclei that are activated by sex steroids to increase the motivation to copulate and, in songbirds, to sing. But where are the neurons that motivate the performance of courtship behavior? Given that male manakin courtship display consists of multiple individual behavioral elements, might we discover that different sites within the brain motivate performance of each specific element? Perhaps the brain possesses premotor neural circuits that take commands from these various motivational centers and modify and relay information to (and from) spinal motor and sensory neurons. Assuming that some or all of these neurons are sensitive to androgens, we returned to the in situ hybridization procedure to assess where in the brains of male manakins androgen receptor neurons were located.

Some of what we found was expected. For example, we did find ample androgen receptor expression in a hypothalamic nucleus called the POM (the medial preoptic nucleus). This has been observed in other bird species and has been identified as a likely site where male copulatory behaviors are activated. We also found androgen receptor mRNA expression in a midbrain region called the ICo, or nucleus intercollicularis. This region is thought to serve a premotor function in regulating calling behavior in birds. Calling behaviors are the unlearned vocalizations produced by most bird species. A rooster's crow is a call. The oscine songbirds both sing and call, as do parrots and hummingbirds. Other birds, including suboscine songbirds like manakins, just call. Such vocalizations may be used during periods of reproduction, for mate attraction or territorial defense, and they are often activated by sex steroids. Thus, the identification of androgen receptors in the manakin ICo was not surprising, as the males do call frequently. Those cheepoo vocalizations that males emit during courtship appear to be activated, at least in part, by androgens acting on the manakin ICo.

What did stand out in the manakin brain was the abundance of androgen-receptor mRNA in the cerebellum, a brain region with known motor control functions. Another brain region rich with androgen receptors was the manakin arcopallium. As discussed previously, the arcopallium houses a key premotor nucleus RA of the oscine songbird song system. Recall that the motor output of song includes the nucleus HVC that projects to RA, which in turn projects to the nXIIts nucleus, where the primary motoneurons that control contractions of the syrinx and thus the output of song are located. The arcopallium contains other premotor neural circuits as well. Could this large field of androgen receptor–expressing cells be where androgens activate the motivation to perform courtship? This seems highly likely in that we have failed to see such a large robust expression of androgen receptors in the arcopallium of birds that do not perform complex courtship. We have not established this connection unequivocally—but we are looking.

It is exciting to think that this region may have primary premotor control over both song in oscine songbirds and aspects of courtship in the suboscine manakins. Song, of course, communicates information to rival males and attentive females. It is acoustic in nature and produced by the syrinx. Because it is a learned behavior, it is quite similar to human speech, and songbirds are studied as a model for human language. While suboscine songbirds do not sing, they do use calls to communicate to other males and females—again, an acoustic form of communication. Golden-collared manakins employ their own unique sort of acoustic communication, in the form of wing- and rollsnaps. Could it be that central (brain) centers involved in acoustic communication project both to the syrinx in songbirds to create song and to the forelimb muscles in manakins to create wingsnaps? Is then wingsnapping somewhat analogous to singing, to language—a sort of noisy sign language?

We will return to this idea in a later chapter. For now, we have identified several manakin brain regions where androgens may increase the motivation to perform courtship displays, as well as regions where androgens may promote the brain's profoundly complex motor output that becomes the male's intricate, skillful, and athletic courtship display.

Earlier I described Leo Fusani's experiment in which courtship behavior was interrupted in wild adult male manakins by the androgen receptor blocker flutamide; although the results pointed to a role for these receptors in courtship, we needed a better way to assess androgen action on androgen receptors. We decided to try a different androgen receptor blocker. This time, together with my previous postdoc and now colleague Matt Fuxjager, and based on a suggestion by Leo, we chose a drug called bicalutamide, also known as Casodex. Interestingly, this drug had been shown previously to not enter the brain. That's right: it can act everywhere in the body *except* the central nervous system. In fact, this is not an uncommon property of many drugs, and it can either be an advantage or cause real headaches in drug design. What's happening is that the brain is protected by what is known as the blood-brain barrier, which shields the very sensitive brain from exposure to potentially harmful agents that might be in the blood. The barrier is created by specializations of the vascular system, primarily the capillaries that ramify throughout the brain; these supply the brain with a copious amount of oxygen, but also provide substantial control for entry of just about anything else. A few molecules get through the blood-brain barrier, such as the lipophylic steroid hormones. Some substances are allowed access to the brain in limited quantities only. Some compounds cannot get through at all—and one of these is Casodex.

We repeated the earlier experiment by giving wild male manakins implants that were either empty (as controls) or filled with Casodex.

The results were clear: Casodex significantly reduced virtually all of the male's courtship display elements. Moreover, the effect not only appeared in the first week after treatment, but persisted into the second week. Overall, the results confirmed that the activation of androgen receptors was crucial for promoting the performance of male courtship displays. Specifically, males treated with a blank implant performed rollsnaps consisting of about eighteen individual snaps, each separated by almost exactly 20 milliseconds. Remarkably, one male performed twenty-three snaps in sequence. By contrast, males treated with Casodex performed only about eight snaps, and the duration between snaps was significantly longer and more variable. In addition, the Casodex trial suggested that the androgen receptors in peripheral tissues, like manakin skeletal muscles, played an important role, such that blocking those receptors almost completely eliminated behavioral performance. Thus, as with athletes taking steroids to build muscle mass in hopes of improving their performance, the testosterone produced, quite naturally, by manakin gonads acts on the androgen receptors in their muscles to make them into acrobats and dancers and percussionists who are also capable of lifting heavy objects to keep their courts clean, all in the hopes of impressing females.

These results raised another question: do manakin skeletal muscles actually express androgen receptors?

We looked, and they do.

Polymerase chain reaction (PCR) is a highly sensitive tool for amplifying minute quantities of DNA to make them detectable for study. Most of us are familiar with the procedure as a way of detecting the presence of a virus, or the amount of Neanderthal DNA we might carry in our genomes. PCR is also useful for exploring the presence of mRNA in extremely small tissue samples, even within individual cells. The procedure involves first converting mRNA back into DNA (complementary DNA or cDNA), which is then amplified in a machine that creates the appropriate environment, including temperature, for strands of DNA to be replicated repeatedly. If a single mRNA (converted to cDNA) is present, and it, and its products, are doubled with each cycle, then millions, even billions, of copies of that cDNA can be produced. In this way, one can observe whether that mRNA exists at all.

One can modify PCR to very accurately quantify the amount of amplification taking place. Just imagine how complicated this amplification becomes if you have not 1 copy of the mRNA but, to pick amounts at random, let's say you have 37 copies to start, or 3,643 copies, or 537,641 copies, and with each step of the PCR reaction, these numbers are doubled. Let's say you want to compare two tissues, one with 537,641 copies, the other with half that many, 268,821. This is likely to have functional implications, with one tissue possibly making twice as much of the protein coded for by the mRNA. Being able to accurately measure this difference is crucial. The procedure to do so is called quantitative PCR (qPCR) and allows for real-time measures of the amplification of the product. If one can draw the curve for amplification over time, the curve that starts with more initial mRNA (and thus cDNA) will rise more quickly than the one that starts with less. By comparing these amplification curves over time, one can accurately assess differences between the two samples. PCR machines can gauge the product at each amplification cycle using special dyes, lasers, and computer programs, and by adding primers, in our case, designed to measure androgen receptor mRNA. Altogether, this procedure can be used to ask if a tissue, such as a muscle, has androgen receptors and how many it has compared to another muscle. One can also ask how androgen receptor abundance compares in a specific muscle between two species. We have done both of these kinds of experiments with golden-collared manakin skeletal muscles as well as manakin heart muscle, testes, and regions of the brain and spinal cord.

What we have found is that golden-collared manakin skeletal muscles are indeed rich with androgen receptors. First, the several skeletal muscles that we examined have a great deal more androgen receptor mRNA than do other tissues that typically are seen as especially rich with these receptors, including the brain, spinal cord, and testes. Moreover, when we made comparisons to two similar-sized passerine birds, captive zebra finches and wild Panamanian ochre-bellied flycatchers, we found that the manakins had up to *twenty times* as much receptor in their muscles. It seems that androgens may exert an inordinate amount of control over manakin muscles.

The muscles that we examined included SH, SC, and PEC, the forelimb muscles that we know are innervated by androgen receptor–expressing motor and sensory neurons. Our results show that the full peripheral neuromuscular system controlling the manakin's wings has sensitivity to androgens, and this is likely responsible for the speed, strength, and precision of the manakin wing- and rollsnaps.

This experiment measuring manakin muscle androgen receptor expression was performed by an undergraduate student in my lab at UCLA, Jenny (Ni) Feng. It is a real privilege to work with bright, curious, dedicated students such as Jenny. After leaving my lab (where she published several papers, including one as first author), Jenny pursued a PhD at Cornell University, did impressive postdoctoral research at Yale University, and is now an assistant professor at Wesleyan University.

While our focus has been on the manakin forelimb muscles because wingsnapping is such a prominent, and unique, feature of male courtship, we also found that other manakin skeletal muscles expressed androgen receptors in similarly large amounts, including the gluteal, a leg muscle. Manakins jump, rather than fly, from perch to perch during courtship, and they hold themselves at awkward horizontal angles while perched on upright saplings. The gluteals (the muscles that form our buttocks) attach the back of the femur to the hip and, among other actions, promote extension of the leg when the knee is bent. Thus, it facilitates jumping and may be especially important to the male manakin in its courtship jumps.

We considered it likely that androgens promote the extra strength of the leg muscles needed for these elements of male courtship. Indeed, a vast number of skeletal muscles in the manakin's body may respond to androgens to promote courtship. Bear in mind, however, that some muscles may need to contract more quickly, or more strongly, or may need to burn fuel more efficiently than others, with androgens playing a role in each of these adaptations. How does one hormone binding to one receptor produce such different properties in one general kind of tissue? We will explore this question in a later chapter.

Interestingly, female manakins were found to have just as much muscle androgen receptor as do males. How and why is that? If females do not perform courtship displays, why do they have so much androgen receptor? The answer is unclear. It may simply be that the evolution of the muscle androgen receptor phenotype is driven by genes unrelated to sex. Thus, all golden-collared manakins express androgen receptors to a high degree in muscle, even if they lack circulating testosterone to activate those receptors.

These observations lead to other considerations. There is some evidence that the amount of androgen receptor in muscle is influenced by how much the muscle is used—that is, if you exercise more, you will have more receptor on which your own testosterone (or a synthetic steroid) can act. If female manakins, which do not perform courtship and do not have elevated levels of testosterone, have just as much receptor as do adult males, which do perform courtship and also have elevated levels of testosterone, then in all likelihood exercise (muscle use) and circulating testosterone have little to do with the amount of receptor in muscle. This suggests that the amount of receptor in golden-collared manakins is a species-specific, constitutive trait.

Female zebra finches also express androgen receptors in their skeletal muscles and at levels similar to male zebra finches. This suggests that androgen receptor expression is species-specific and constitutive in zebra finches as well.

Golden-collared manakins have an elaborate courtship display with lots of androgen receptors in their skeletal muscles. Both zebra finches and ochre-bellied flycatchers have minimal courtship displays, though the flycatchers do use their wings somewhat in courtship. This begs the question of whether the amount of androgen receptor present in skeletal muscles is associated with courtship. Put another way, is the evolution of elevated androgen receptor expression in muscle a prerequisite adaptation enabling circulating androgens to promote the subsequent muscle-specific states required for elaborate physical courtship displays?

Keeping in mind that there is ample androgen receptor in the spinal cords of manakins, it might be the case that the amount of androgen receptor in the spinal cord predicts display complexity. We had already found more receptor in the spinal cords of manakins than in zebra finches, and potentially more in male manakins than in female manakins. So perhaps elevated expression of androgen receptors in both muscle and neural systems preadapts any bird species for the performance of complex androgen-dependent courtship?

When Matt Fuxjager was still a postdoctoral fellow in my lab, he tested these ideas in collaboration with other manakin researchers. The question asked is, does the amount of androgen receptor in muscle or spinal cord correlate with the complexity of the bird's courtship behavior? Of course, how do you quantify complexity to make comparisons when species differ in the kinds of behaviors they incorporate in their displays? This question was addressed first in manakins by Rick Prum. Recall that, together with Kim Bostwick, Rick first published high-speed video of golden-collared manakin wingsnaps. In addition, he has over the years studied and described the courtship displays of a variety of different manakin (and non-manakin) species. He created a scoring system that attempted to quantify the overall physicality of the displays across species, including the amounts and types of postural, wing, leg, neck, and head movements. We adapted Prum's system for our approach.

We chose to compare a group of small passerine birds that differed widely in their displays; these included manakin and nonmanakin species as well as oscine and suboscine passeriform birds. All of the birds were captured in the wild except the zebra finches, which came from our captive colony at UCLA. The manakin species we chose (golden-collared, red-capped, blue-crowned, lance-tailed) were all collected in Panama, as were the ochre-bellied flycatchers. We also included pin-tailed whydahs (*Vidua macroura*), small finches much like zebra finches; though native to Africa, they have become resident on the island of Puerto Rico, where they were collected by Lucie Salwiczek, who had joined my lab as a postdoctoral fellow from Germany. For each of these species, we computed two scores, one that focused on the complexity of the wing movements within a display and another that described overall display complexity. To generate scores, we assigned a value of 1 if a behavior was present and 0 if absent. Thus, in this system, higher scores correspond to more complex displays and lower scores correspond to less complex displays. Both golden-collared and red-capped manakin males had the most complex total physical displays (with a score of 18 each), followed by lance-tailed manakins (17), blue-crowned manakins (11), ochre-bellied flycatchers (7), and pin-tailed whydahs (4), with zebra finches bringing up the rear (score of 2). Zebra finches have no wing component to their displays (scoring 0), while lance-tailed and red-capped manakins have the most (scoring 12), with the golden-collared manakin close behind (11).

We obtained forelimb muscle and spinal cord tissues for these experiments, including the SH, SC, and PEC muscles and the spinal cord dissected into rostral and caudal sections. Once back in our lab at UCLA, we extracted mRNA from all of these tissues, made cDNA, optimized all procedures and reagents for all species, and then measured the amount of androgen receptor present in each tissue. In addition, and as a control, we measured the amount of estrogen receptor. We did not expect the amount of estrogen receptor to relate to much of anything, because we have minimal evidence that estrogens play a role in male courtship.

What we found was striking. First, we were able to detect both androgen receptor and estrogen receptor expression in muscles and spinal cords of all species. However, when we related estrogen receptor expression in muscles as well as estrogen and androgen receptor expression in spinal cords, we found no relationship to courtship complexity. By contrast, the amount of androgen receptor in skeletal muscles was *highly* correlated with both overall motor complexity and wing movement complexity. (Note: we only examined receptor expression in a subset of forelimb muscles.) All of the bird species line up fairly closely along a central straight line connecting the zebra finch, which has minimal physical courtship and little androgen receptor expression, to the golden-collared manakin, which has a lot of both, indicating that the two variables are tightly linked traits.

Another way scientists think of this is by considering the data's R^2 value. Imagine all the birds lined up perfectly on the straight line connecting the finch and the manakin. That would indicate a 100 percent relationship between display complexity and androgen receptor expression, or an R^2 value of 1.0. One could conclude that the amount of androgen receptor in muscle completely predicts the display complexity score. Our calculations produced an R^2 value of 0.76—that is, muscle androgen receptor levels predicted 76 percent of the total display complexity score. When you consider all the different factors that might influence display complexity—the function of the brain, the spinal cord, and other muscles and neural circuits, along with all of the various proteins and myofiber types contained within the muscles—an R^2 value of 0.76 is very high.

The importance of androgen receptors in skeletal muscles to promote athleticism in bird courtship displays cannot be overestimated. But just what do these receptors do? What happens when testosterone activates these receptors in muscles? Are all of the actions the same, or do these actions differ from muscle to muscle? We explore these questions in the next chapter.

TWELVE

Manakin Muscles Are True to the Task

Knowing that androgens have such powerful effects on male manakin biology is, at some level, not terribly surprising. We all have some notion about male hormones promoting masculine behaviors. One way that the manakin case is special is that we can explore just how androgens make a muscle better. On top of this, we can link this muscle action to the larger scope of manakin life history and behavioral ecology, one of the goals of this book. What is still needed is deeper inspection of the molecular changes that are happening in a contracting and relaxing muscle and just what androgens can do to influence those different muscle states.

But first we have to make a small leap of faith. That is, we have to assume that the mRNA we detect by in situ hybridization or by PCR is indeed translated into protein—that it is a real androgen receptor. Further, we have to assume that all the other cellular ingredients exist to allow androgens to bind to that receptor, and that they affect changes in the cells that possess that receptor. While we have confidence that these steps do indeed occur, at some point these processes should be confirmed. For now, though, the fact that we can block androgen action on a cell's receptors and effect changes in behavior gives us license to ask just what androgen is doing to these cells and tissues.

One way to answer this question is to focus on the contraction of individual muscles, as opposed to considering courtship behavior in its entirety. Recall that the wingsnapper's SH muscle contracts at the fastest rate of any vertebrate limb muscle. Further, it is important to realize that in order to contract repeatedly and rapidly, the muscle must also relax quickly. To test this, Matt Fuxjager and Franz Goller examined the twitch speed of SH muscles in male manakins treated with testosterone or testosterone plus flutamide, the drug that blocks androgen receptors. They discovered that, indeed, androgen activation of the receptors converted the SH muscle from a fairly mundane wing muscle into one with an incredibly fast relaxation speed; that is, the muscle could quickly rebound after one contraction to fully contract again, just what is needed to create rollsnaps. Remarkably, this increase in contraction speed occurred while the muscle maintained its contraction strength. All in all, an SH muscle on androgen is fast and powerful.

These results led us next to ask just what genes are expressed in the androgen-activated SH muscle that make them so fast and powerful. Are there genes known to be expressed in skeletal muscles of other species also expressed in manakin muscles but now under androgen control? Might manakins have evolved newer protein forms that enhance their muscle contractility? In studying gene expression in muscle, our primary interest was the forelimb muscles that are involved in producing wingsnaps, so we examined the SH, SC and PEC, as we had done previously. Because we had found that skeletal muscles of zebra finches also had androgen receptors, but much less so than manakins, we asked if the same muscles in zebra finches also expressed these same genes and if the regulation of these genes occurred to a greater degree in manakins than in zebra finches. For these studies, male manakins were captured during the nonbreeding season, so the birds naturally had low circulating testosterone levels (like castrates). Half of these manakins were given an implant of testosterone and the other half a blank implant. All of the zebra finches were treated with testosterone, but half of these were also treated with the androgen receptor blocker flutamide.

We selected four genes for study. One of those genes produces a molecule called parvalbumin, which serves as a buffer to intracellular calcium concentrations. As you might remember, calcium is the trigger for muscle contractile proteins to shift their configuration, which in turn generates muscle contractions. For a single wingsnap, calcium is released, the wing muscle contracts, and a wingsnap is produced. But what about a rollsnap, a very rapid series of individual wingsnaps? In this case, calcium is released with each individual snap, but it must then be quickly removed so the muscle can relax, allowing the next release of calcium and the next contraction and snap. For a good long rollsnap, this contraction/relaxation cycle is repeated eighteen or more times. The removal of calcium between individual snaps involves buffering, and that is where parvalbumin comes in. So, one would predict that any muscle that contracts multiple times rapidly, like the forelimb muscles of male manakins, might contain lots of parvalbumin.

We also wondered if androgens influence muscle growth or size. We examined three genes: IGF-1 is a well-studied growth factor that is known to enhance muscle size and strength by inducing muscle cell proliferation and enlargement; MyoD increases muscle strength and size by regulating the production of proteins that affect structure and fiber-type composition; myostatin negatively regulates (i.e. prevents) muscle growth by arresting cellular division of muscle cell precursors. (There is a famous journal cover-photo of an Arnold Schwarzenegger–like mouse that was created when transcription of the gene for myostatin was eliminated from its skeletal muscles.) Thus, we predicted that IGF-1 and MyoD would be expressed in greater amounts in manakin muscles than in zebra finch muscles, and that each of these genes might also be upregulated by androgens, especially in manakins. Because myostatin reduces muscle size, we predicted that this gene would be expressed at low levels in manakin muscles and perhaps even be reduced by androgens.

As a first pass, we found that all of these genes were indeed expressed in at least some muscles from these birds. Beyond that, some of our results were anticipated, others not. For example, parvalbumin and IGF-1 were generally expressed at higher levels in manakins than in zebra finches, as expected. Surprisingly, however, myostatin was expressed at relatively high levels in some manakin muscles as opposed to zebra finches, including the SH muscle, which is quite bulky in manakins and quite puny in zebra finches. Clearly, low expression of myostatin does not, at least by itself, make male manakins buff.

Our data also showed that androgens could have a strong impact on gene expression in bird muscles, especially in manakins. Parvalbumin expression illustrates this clearly. First, levels of parvalbumin were significantly higher in all muscles from manakins treated with testosterone as compared to blank-treated manakins. A similar androgen dependence was seen in zebra finches. There also seemed to be a relationship between the amount of androgen receptor and the amount of parvalbumin expressed: manakins with more muscle androgen receptor had more parvalbumin, while zebra finches' muscles with less androgen receptor had less parvalbumin.

The results for IGF-1 were similar to those of parvalbumin. Thus, a growth factor that stimulates size and strength of skeletal muscle is upregulated in muscle by androgen with the increase greater in muscles with more androgen receptor present—that is, more in manakins than in zebra finches.

In summary, androgens promote expression of IGF-1 as well as the calcium buffering molecule parvalbumin, which together make the SH muscle large and strong and capable of relaxing quickly. One result of having lots of these two molecules is the male's remarkable rollsnap behavior. The data for MyoD and myostatin were less conclusive, suggesting that they might not be responsive to androgens in these species. Importantly, the results of these studies highlight the fact that functional androgen receptors exist in avian wing muscles.

Interestingly, this study of manakin wing muscles gives us some hints about what might be going on in muscles of other bird species. Consider birds that migrate. To engage in incredibly difficult long-distance migratory flight, birds modify many tissues in their bodies. For example, because birds do not use their legs during migration but flap their wings an extraordinary number of times, they expand the sizes of some of their wing muscles and they reduce the sizes of some of their leg muscles. One such migratory bird is the white-crowned sparrow (*Zonotrichia leucophrys*) of North America, in which the potential role for androgens in sculpting their migration has been studied. As it turns out, androgens upregulate IGF-1 in a white-crowned sparrow wing muscle, but *reduce* IGF-1 in a leg muscle, and this occurs due to androgen action when the birds are preparing to migrate north in the springtime. Thus, androgen control of IGF-1 seems an important property of avian muscles, with androgen having the capacity to greatly increase *or* decrease the presence of IGF-1 depending on the muscle. How the same gene in different muscles responds differently to androgens is unknown, but experiments are underway to understand the mechanism.

Although our experiments with these few genes give us guidance for what makes muscles function properly in courtship (and migration), we have a long way to go. The avian genome has tens of thousands of genes, and any specific cell in any particular tissue expresses thousands of genes at any one time. There is, thus, the potential for androgens, acting through androgen receptors, to influence the expression of a very large number of genes. But how does one investigate a question like this?

■ ■ ■

We now have the technical capability to collect all of the messenger RNAs from a single tissue and to have them sequenced and studied. All of these mRNAs are created by transcription of the genome; hence this collection of mRNAs is called the transcriptome. Because each tissue in the body performs unique functions, each tissue has its own characteristic transcriptome; similar tissues share a great deal of the transcriptome, while very different tissues share much less. Thus, the expressed genes that produce a functional kidney cell will differ markedly from the expressed genes that produce a functional neuron. Two skeletal muscles will share a large number of transcripts that dictate basic muscle function. Yet in that most muscles have slightly different characteristics—traits like fast-twitch or slow-twitch, for example, which derive from the presence of unique sets of contractile and/or metabolic proteins—those transcriptomes will reflect those differences.

Some of what makes the transcriptomes differ between tissues involves the identity of the genes that are transcribed, whereas some of the ways tissues differ is in the number of transcripts that are present.

There is an enormous amount of information in a transcriptome, and unraveling all the mysteries inherent in these data requires advanced and specialized computer programming. This is the business of the scientific domain called bioinformatics. In addition to possessing remarkable intellectual skills, bioinformaticians must exercise great patience, especially when dealing with individuals like myself, who, working with little money on unique tissues from novel species, will have an abundance of curiosity and exuberance and ask many, many questions, including, hopefully, some good ones.

Several years ago, our department hired just such a computational biologist, Xinshu (Grace) Xiao. Now a full professor, Grace is in constant demand for collaborations, and she serves on many committees and programs that require bioinformatics expertise. For some reason, she accepted my request to collaborate on manakins, though I'm not sure she knew what she was getting herself into. Still, she agreed with a smile, and seems to be smiling to this day.

Matt Fuxjager and I provided Grace with wing muscles from a golden-collared manakin and a zebra finch that were either androgen-activated (i.e. treated with testosterone) or nonhormonally activated (i.e. treated with a drug to block androgen receptors and any actions they might promote). Grace's lab extracted RNA and sent the samples off for sequencing. The result was millions upon millions of transcripts that required extensive bioinformatics analysis.

We learned a lot from this work. One thing we learned is that the same muscles in both zebra finches and manakins are often quite different from one another. We also learned that, especially in manakins, specific individual muscles could differ greatly within a single species. For example, the pectoralis muscle that lowers the wing expresses different genes than does the SH muscle, which produces the wingsnap. It was fascinating to see just how different two skeletal muscles can be. Most striking of all was that androgens affected manakin muscles far more than they did zebra finch muscles, with the wingsnapping SH muscle impacted the most. For example, not unexpectedly, androgens regulated many genes that increased the speed and strength of muscle contraction, but androgens also regulated genes that regulated manakin muscle energy efficiency. Clearly, manakin muscles are unique, and androgens make them all the more fascinating.

Let us step back to consider all of this muscle biology from a broader perspective, indeed from the standpoint of any organism with muscles. As you move about, think of all the muscles that propel you, and consider just how each muscle has its own personality with regard to contraction rates and duration or the amount of energy it consumes to achieve its movements. All of these specializations arise from the unique expression of genes in each of those muscles. Every bird that flies overhead, too, is using numerous muscles, all with their own personalities. Think of all the muscles of birds that engage in elaborate courtship using an array of adaptations driven by the suites of genes that have evolved to promote proper muscle function.

Remember that each of these muscles is innervated by neural circuits that require activation of motivational and premotor systems within the brain that transmit an extraordinary array of rapid-fire signals down the spinal cord to motor and sensory systems that communicate with the musculature. This is extraordinary biology. In manakins, all of these neural and muscular specializations yield the remarkable set of behaviors that we witness in their courtship dance.

PART III

A Spotlight on Female Manakins—and Us

THIRTEEN

Females Shape Male Manakin Behavior

Although the bulk of this book is focused on the behaviors and adaptations of male golden-collared manakins, female manakins are truly the driving force behind male manakin behavioral ecology and evolution.

For a good part of the year, from June through January, females, like males, spend their days searching for food, avoiding predators, dealing with ticks, mites, and mosquitos, and staying as dry as possible in the very wet and sometimes stormy tropical rainforest. From February through May, however, their lives are entirely different. This is the breeding season, and female physiology changes markedly. The reproductive centers of their brains become activated, likely triggered by the small increases in day length that occur at this time of year (on the order of minutes), though they may also be responding to an improved diet due to the reduction in rainfall and the emergence of small fruits and insects.

The brain and the ovary are linked in remarkable ways. Cells in the hypothalamus begin to synthesize and secrete a hormone called gonadotropin-releasing hormone (GnRH). (It is also sometimes called luteinizing hormone-releasing hormone, LHRH, but this term is biased to mammals and we will not use it here.) When sensory information tells the brain that the time is right to breed, GnRH is released into capillaries that collect together and exit the brain, traveling a very short distance to the pituitary gland, which sits just beneath the brain. The anterior part of the pituitary has cells called gonadotropes, as well as many other cells crucial for a variety of physiological systems. When GnRH from the brain hits gonadotropes in the pituitary, they are stimulated to release two hormones into the general circulation, LH, luteinizing hormone, and FSH, follicle-stimulating hormone. This occurs in both sexes, but in females these two hormones stimulate the ovaries to grow. Ovarian follicles swell and secrete copious amounts of hormones, first estradiol, followed somewhat later by progesterone.

These two steroid hormones act, respectively, on estrogen receptors and progesterone receptors, producing profound changes in the anatomy, physiology, and behavior of the female manakins. Estradiol takes the lead and stimulates all the viable reproductive systems, including the regions of the brain that make females attracted to males and receptive to their acoustic and visual signals. Upon receipt of a satisfactory signal, estradiol makes the females prone to assuming positions that allow males to copulate. It causes the oviducts to enlarge substantially and to produce biochemical changes in the cells lining the oviducts. It also causes the liver to produce unique sets of lipids and proteins. The ovarian follicles, each with a single ovum ready to be fertilized, begin to swell as estrogens instruct those ova to capture the circulating lipoproteins from the blood. Each species grows a specific number of ova. A chicken, for example, can lay an egg each day for several days, and then it takes a breather. These eggs are called the bird's clutch. Some wild ducks and galliform birds like quail can produce a clutch of up to twenty eggs. After hatching, these precocial ducklings or chicks can be seen chasing after the mother in a long line. Enormous condors have a clutch of one. Tiny manakins have a clutch of two.

This brief description of female reproduction in birds belies the fact that it is an extraordinary biological accomplishment, so beautifully intricate in its orchestration.

Probably stimulated by estradiol, the female manakin now finds a suitable place to build a nest, then goes about all of the intricate motor patterns to collect and assemble material into a cup that will fit her eggs and her body.

I, for one, find nest building in animals, especially birds, to be an underappreciated behavior, one that leads to true architectural feats. A human must be intensively trained to create something as beautiful and well-constructed as the nests of many bird species, who do the weaving primarily with their beaks, using material they find in nature. The African weaver-birds are especially well known for the skill they demonstrate in building nests, but birds of all types make remarkable nests.

Nests of birds in the tropics are often the most complex, likely due to the intensity of the environmental conditions they must withstand to successfully raise their broods. The tropics can be very hot and wet, with storms that can bring fierce winds. And then there is nest predation. Just about everything eats eggs and nestlings—snakes and monkeys and, in the neotropics, long-nosed coatis. Swarming army ants will devour anything edible in their path that cannot escape, including hapless nestlings. Other birds eat eggs and young as well. Toucans, for example, use their elongated, colorful beaks to pluck defenseless little nestlings from exposed nests. Even cute little wrens, with their impressive songs, use their long, pointy beaks to pierce eggs and carry them off.

Near where golden-collared manakins live are several species of caciques and oropendolas, birds related to the American blackbirds and orioles, but generally larger. These impressive birds build aston-

Fig 13.1. Nest with two eggs of a golden-collared manakin.
Reproduced with permission from Frank M. Chapman,
"The courtship of Gould's manakin (*Manacus vitellinus vitellinus*)
on Barro Colorado Island, Canal Zone," *American Museum of
Natural History Bulletin* 68 (1935): 472–521.

ishing basket nests up to two meters long, which are woven to hang
from the outer branches of tall trees. These birds nest colonially, so
a tree, perhaps 30 meters or more tall, may be laden with dozens of
these large penduline nests, like huge hanging fruits. Such trees are
a characteristic feature of the lowlands of many neotropical forests.
A nest like this, sitting far from the center of the tree, is hard for
most predators to reach. Those that do try will be attacked by a score
of noisy guardian birds, making nest-robbing success doubtful.

In some species, both males and females participate in nest building. Not in golden-collared manakins. With males too busy dancing to help, the task of building a nest is the female's alone. Even
though most of these nests are built relatively close to leks and are
situated somewhat out in the open, they can be quite difficult to spot.

The female selects a thin upright sapling, a meter or so tall, from which several branchlets emerge laterally. The ones I have found are leafless. I do not know if females choose these because of their sparse cover, or if they actively remove the leaves, like males do from their arenas, when they interfere with the creation of a nest. In any event, the female weaves her small cup-like basket nest in between the little branchlets so it sits directly above the upright stem (fig. 13.1). Here the nest is quite secure and surprisingly well camouflaged, just dead-looking vegetable matter on a nondescript small shrub. Eventually, the female will deposit two brown-spotted and -streaked eggs into the cup and then sit for much of the day to incubate.

Frank Chapman found very few golden-collared manakin nests himself, but relied on the observations of others, including the now-famous ornithologists Alfred O. Gross, Jocelyn Van Tyne, and Alexander Skutch. Those reports from nearly a hundred years ago remain informative today. Here is Chapman:

> Dr. Gross writes of three nests found near the mouth of the brook that parallels the Lutz Trail and enters the lake at the easterly side of the Laboratory dock. From his data I select the following:
> "No. 1. Found July 10, 1925, by F. Drayton; contained two eggs. The bird remained incubating until I was within two feet of the nest. July 22, 1930, A.M. One egg hatched. July 24. Nest and young destroyed during the night.
> No. 2. Found July 22, near No. 1, about four feet from the ground; two eggs."
> Dr. Van Tyne also studied nest No. 1 and, in addition, he reports the discovery on July 12, 1925, of a nest about "200 feet farther up the brook" and on August 1, 1925, of one "in the forest just west of the laboratory." The first contained eggs which hatched on the 14th. It was under daily observation until July 18 when it was found destroyed. The second contained

one egg on August 1, two on the 3d. They hatched on the 22nd, after an incubation period, therefore, of nineteen days. Dr. Van Tyne also records the taking of an incubating female on August 26.

Both these observers state that males were neither seen nor heard near the nests observed by them.

In 1935, Dr. Alexander Skutch discovered four nests and contributes the following data concerning them. In no instance was a male seen near them.

"February 27. To-day I found a recently begun nest of this species, in the horizontal fork of a small sapling growing in a rather open space in the woods. It is only 20 inches above the ground, and quite near some "courts" of the males (40 ft. from the nearest).

March 4. The nest appears to be finished.

March 16. Still no eggs. Apparently abandoned.

March 16. Nest 31 inches above the ground in a horizontal fork of a sapling growing in second-growth forest, near the same "courts" as No. 1, but on the other side of the trail and somewhat more distant.

It is a shallow, open cup of brown fibres, the meshwork so loose that the ground may be seen through the bottom, attached at its margin by cobweb and some fibres passed over the arms of the fork. It measures 2 inches in internal diameter and 1.1 inches deep. 1 egg, whitish, very heavily marked with brown, chiefly in the form of irregular longitudinal marks on the sides.

March 19. The egg has vanished. I never saw a bird on the nest, for no matter how carefully I approached, the nest would be vacant when I arrived within sight of it; but the warm egg told that she had just flown. The identification of this nest depends upon its similarity to nest 1, and the description of the eggs given me by Dr. Chapman."

Dr. Skutch also found nests on April 17 and May 7, 1935, both about 300 yards from the Donato courts. The first contained a single egg on the 19th, which was present on the 20th, but on the 21st it was missing. The second contained one egg when found and this number had not been increased on the 13th. On the 14th the nest was empty.

Dr. Skutch writes that the males were still "snapping freely" when he left Barro Colorado at the beginning of June.

Mr. Harrower's notes from Gatun, given below, include a record of the distance from the nest of courting males and also the leaving of the nest by the young:

"Nest 1. July 23, 1933. Below the dam at Gatun. Nest hung in the fork of a small tree at the edge of a partial clearing in second-growth forest, six feet up. A shallow cup resembling the nest of the Acadian flycatcher, composed of plant fibres, fine strips of material like bark, grass, etc., lining scarcely finer. Eggs 2, cream heavily blotched and streaked with different shades of rich brown tending towards purple. Apparently fresh.

The female flushed from the nest when approached within a hundred feet and flew low into the neighboring jungle. She was so shy that I was compelled to make several attempts at identification before I was certain of her identity, though I suspected it on finding the nest. During the three weeks in which I watched this nest I never once observed the male in the vicinity, though I made no special study of it.

Even with the young hatched the female remained surprisingly shy, and disappeared completely when the nest was approached.

About two hundred feet away was a small thicket in the center of the clearing where several males congregated at times and went through their characteristic antics and displays.
July 29. Two small young.
July 31. Young OK.

Aug. 5. Young partially feathered.

Aug. 7. Young crowding in nest.

Aug. 12. Young gone.

Nest 2. July 27, 1933. Along "Hill Trail" up Gatun Hill beyond dam. Nest hung in the fork of a small bush along trail, four feet above the ground. In no detail of importance differing from Number 1. Contained two rather well-fledged nestlings.

Female did not appear to be nearly as shy as Number 1, and remained close at hand in the jungle while I was at the nest. She hopped about in the low growth in a manner that at times suggested a warbler.

July 29. Young nearly ready to leave nest.

Aug. 2. Nest empty.

Nest 3. August 17, 1933. On the top of a small ridge running out to top of Gatun dam across spillway. In dense second-growth jungle, not at edge of clearing or trail. Nest in fork of small bush, about three feet from ground. Contained two fully fledged nestlings which flew out of nest at my approach and vanished in the low jungle.

Female was again in evidence, scolding from thicket. As I recall it, she uttered one note."

[Chapman concluded:] The facts pertinent to the subject of this paper contained in the preceding observations are (1) the extension of the nesting season of Gould's manakin to at least late August, and of the courtship season to at least August 12; (2) that the nest may be as near as forty feet to an occupied court; (3) that the male takes no part in nest life.[1]

At some time prior to laying her eggs, before she ovulates her yolky ovum, the female must find a male and allow him to copulate and transfer his sperm to her cloaca. Estradiol acts on neural circuits in the female's brain to increase her motivation in this regard. When that estradiol is matched with the sensory stimulation that says, "This is the right male," the female holds still, swishes her tail a bit, and allows the male to land on her back and touch his cloaca to hers—the cloacal kiss. This brief moment of contact is sufficient to allow for the passage of sperm. The two birds then quickly shake their ruffled feathers, and the female departs, while the male's attention returns to his arena and the wingsnaps of the other males in the lek, just in case another female happens by. Meanwhile, the male's sperm finds its way through her shell gland (or uterus) and up the oviduct to fertilize that yolky ovum before all the ovalbumin is secreted and the shell imprisons the fertilized egg and its nutritious liquid medium.

This act of choice by the female is crucial to much of what is described in the rest of this book. Just what does she experience as she moves through the forest and hears the sounds of wing- and rollsnaps and, perhaps, cheepoos in the distance? Why does she approach a lek of males? How does she feel when suddenly males come out of nowhere and start chasing her, making more snaps and cheepoos? Why does she peer onto the cleared arenas of males and watch them go about their jump-snap displays? What does she see as the males bound about at incredible speed? Why does she pick some males over others to join in their dance? And why does she allow only a subset of these males to copulate?

Unfortunately, we have few answers to these questions. But we can speculate.

First, we know from studies of other species that estrogens promote the copulatory postures that females display at the onset of reproduction. Reproductive behaviors are often described as either appetitive or consummatory. Consummatory behaviors in birds and mammals are those at the climax of the long sequence that leads to successful internal fertilization. For the female manakin, that means the moment she freezes on a sapling of the male's arena and allows the male to land on her back, when she arches her back and moves her tail to allow for the cloacal kiss. Virtually all of the other behaviors prior to this moment, in which she responds to the sensory input from the male's displays, are appetitive.

In all likelihood, shortly after copulating, the female ovulates, the ovum is fertilized, and, as we have seen, estradiol tells the female's liver to secrete special lipids. Estradiol then tells the ovum to collect those lipids from the blood and deposit them around the fertilized ovum; estradiol further tells the oviduct to secrete ovalbumin, and the female's bones to release some of their calcium and other minerals, and the female's shell gland to capture those minerals and secrete them onto the ovalbumin, together with some pigments to camouflage the egg. And then, usually early the next morning, the egg is laid. In many species, one copulation is enough to provide the sperm for the fertilization of one or more ova, though some females copulate daily. In manakins, once two eggs are sitting comfortably in the nest, the female starts to incubate.

As this description makes clear, estradiol promotes most of the anatomy and physiology that allow for successful ovulation and successful fertilization. This is much the same in mammals, including in humans. The next phases of female reproduction are largely controlled by two different hormones, progesterone and prolactin, which trigger maternal physiology and behavior. While birds differ from most mammals by laying eggs, these hormones (including others like mesotocin, the bird version of our oxytocin) play similar roles. Rather than caring for the developing embryo and fetus internally, the female bird incubates her eggs and broods the developing young externally, keeping her hatchlings warm and feeding them until they fledge. Progesterone and prolactin are responsible for the anatomical, physiological, and behavioral changes associated with maternal care. To keep eggs warm and moist during incubation, for example, females of many bird species form a brood patch: the skin of the abdomen loses any feathers it might possess and it becomes highly vascularized and accumulates fluid. In the absence of insulating feathers, the mother's heat can pass through this modified skin patch to keep the eggs at the right temperature for embryonic development to occur.

Almost immediately upon hatching, the tiny, poorly developed altricial nestlings begin to beg. During fetal life, their neck, tongue, and beak muscles, and the brain regions that control them, have developed ahead of other systems, so virtually all the hatchling can do is hold its head up and back, open its beak, wag its tongue, and swallow whatever mom puts there. Adult golden-collared manakins eat much fruit, but this is a poor diet for a rapidly growing nestling. They need protein and other nutrients, so a female manakin probably captures many insects to feed her young, though we seldom see this occur in the wild.

We do not know how long the fledglings remain with the female before they gain independence. What we do know is that in or around June, lots of green birds are suddenly seen in and around the leks. They appear in mistnets placed in the forest by ornithologists studying other species. Trees bearing small fruits are now filled with small forest birds—tanagers, flycatchers, honeycreepers, and, in midsummer, lots of green manakins, often of several species, including many with bright red legs, golden-collared manakins. We do not know the sex or age of these birds, but we suspect that only some are adult females, while most are juveniles, the successful crop derived from all of the efforts of adult females from the preceding breeding season.

The strange set of behaviors that male manakins perform to attract females is amazing to our eyes. But just what do the females see? Although we cannot answer that question for sure, we can make some assumptions. Recall that, in order to see the minute details and full elegance of the male's courtship display, we humans need to use specialized high-speed camera equipment so that we can play back the video in slow motion. We must assume, however, that female manakins can see these details just fine. One reason we reach this conclusion is that males that perform elements of courtship more quickly get more copulations, as if females prefer the speediest males—and here we're talking milliseconds.

But why would speed be important to a female? Several manakin species move quickly as they go about their normal business—when flitting from perch to perch, for example. Now you see them, now you don't. A moderately talented birdwatcher can anticipate and follow the movements of many birds, but the manakins are tough. Why so quick?

As we have seen, the rainforest is filled with predators, and they eat more than eggs and nestlings; they also eat adult birds. Think of a vine snake, curled inconspicuously among tangled branches in dense forest. A small bird lands nearby, and in a flash, the bird is caught. Unless, that is, the bird has super-fast visual processing and motor reflexes. Such skills might, of course, be acquired via natural selection. Another evolutionary path for acquiring physical skill might proceed through mate choice, whereby females select the fastest, most talented males, and only those females with the fastest visual processing see the male's speed and grace. The offspring of such matings will inherit the best of both parents. I find it impressive that male manakins live such long lives when they are so conspicuous, both visually and acoustically, and when they occupy the same place—their arena—at pretty much the same times day after day for months on end. Shouldn't a predator figure out that a meal (or a whole lek thereof) will be available to them?

Predators are fast, especially predatory birds. Most of us are familiar with the great speed of falcons. The peregrine (*Falco peregrinus*) is known as the world's fastest animal, reaching speeds of 300 kilometers per hour when diving. They achieve this, though, in midair over open country, far from where manakins live. Their relatives the forest falcons of the American tropics (*Micrastur* spp.), however, occupy the same forests as many manakins, and they navigate this environment easily. Another group of bird predators is the accipiters, the real masters of hunting in thick forest planetwide. The forests of Panama, for example, house the tiny hawk (*Accipiter superciliosus*), which has a taste for hummingbirds. There is also the bicolored hawk (*Accipiter bicolor*), though they are rarely seen in central Panama. Accipiters are fierce and, with their relatively short, rounded wings, are designed for exceptional speed over a short distance, while their relatively long tails allow for quick and accurate maneuverability.

Although they do not live in tropical rainforests, the goshawk of northern forests is the pinnacle accipiter, being the largest and considered the most powerful of this bird-hunting group. I was once attacked by a large female goshawk in central Massachusetts. My companion and I saw her from about 30 meters away, sitting next to her nest. Like a lightning bolt, the bird dove at us; we hit the ground simultaneously as the bird passed over us by mere centimeters. Forest birds don't stand a chance against these skilled hunters.

You may be familiar with one particular goshawk, Mabel, the star of Helen Macdonald's compelling book *H Is for Hawk*, about falconry, death, and life. Macdonald's take on Mabel's perspective is worth quoting:

> The world she lives in is not mine. Life is faster for her; time runs slower. Her eyes can follow the wingbeats of a bee as easily as ours follow the wingbeats of a bird. *What is she seeing?* I wonder, and my brain does backflips trying to imagine it, because I can't. I have three different receptor-sensitivities in my eyes: red, green and blue. Hawks, like other birds, have four. This hawk can see colours I cannot, right into their ultraviolet spectrum. She can see polarized light, too, watch thermals of warm air rise, roil, and spill into clouds, and trace, too, the magnetic lines of force that stretch across the earth. The

light falling into her deep black pupils is registered with such frightening precision that she can see with fierce clarity things I cannot possibly resolve from the generalized blur. The claws on the toes of the house martins overhead. The veins on the wings of the white butterfly hunting its wavering course over the mustards at the end of the garden. I'm standing there, my sorry human eyes overwhelmed by light and detail, while the hawk watches everything with the greedy intensity of a child filling a colouring book, scribbling joyously, blocking in colour, making the pages its own.[2]

To avoid such a powerful and skillful predator, birds develop unique adaptations, including, in some cases, great sensory and motor speed. This is true of many of the manakins.

But in the manakins, as we have seen, the quickest males generally also perform more displays overall, calling more and making more rollsnaps. Presumably, the female pays attention to these individual behaviors, which combine to give the stimulation that she seeks. Perhaps she does not discern each specific movement, but she may keep track of the male's total number of moves and be aware of elapsed time. Or maybe she is paying attention to his wingsnaps, their amplitude or rhythm. My guess is that she is paying attention to all of it.

The female's visual system does seem to be key. At the very least, there is evidence that females pay attention to color. Al Uy and his student Adam Stein conducted a series of studies in a region of western Panama near the border with Costa Rica where the golden-collared manakin hybridizes with the white-collared manakin (*Manacus candei*). This hybrid zone had been studied extensively by Mike Braun and his colleagues at the Smithsonian Institute in Washington, DC, who found that the "golden" collar (or should I say the genes encoding the golden collar) of *M. vitellinus* were introgressing into the white collar of the *candei* individuals. That is, some individuals with a yellow collar were actually genetically more *candei* than *vitellinus*. Along with these genetic results, morphological studies showed that within this hybrid zone, many males with yellow collars possessed phenotypic traits that were more similar to those of the white-collared variety. While hybridization may occur for various reasons, presumably females of one species or the other (which are largely phenotypically indistinguishable across species) prefer to mate with males of the other species

Indeed, according to studies of Uy and Stein, this hybridization occurs because white-collared females prefer to mate with golden-collared males. The researchers focused their studies on a region of the hybrid zone along the spectacular Rio Changuinola, where both white- and yellow-collared males display side by side. They recorded the number of females that visited the leks and the number of copulations observed. Males of both collar colors received about the same number of visits by females, but the yellow-collared males obtained more copulations. Did the females prefer yellow necks over white?

It turned out to be more complicated, because this preference only applied when males with yellow collars were more numerous at the lek than those with white. Were the females keeping count? Probably not (though that would be another interesting question to explore). Uy and Stein surmised that for color to be an effective signal to females, it needed to stand out from the background colors of the forest. It has been suggested that one reason males clear their arenas of debris is to allow better observation of their plumage. Perhaps it is not just the color that is important, but how the plumage contrasts against the surrounding foliage. By measuring the reflectance of the male's plumage, as well as that of the forest surrounding the leks, Uy's group found that, as predicted, the golden collars stood out better in forest occupied by the golden males, while the white collars were more conspicuous in forest occupied by the white males. The foliage of the hybrid zone seemed to favor gold over white.

One conclusion we might draw from all this is that when being courted, the female uses her visual system to assess the opportunities, and the color of the male's collar, especially one that stands out in the forest, is a key piece of information informing her choices.

Recall that during courtship, the male actively presents his "beard," the feathers under his lower mandible. Throughout his courtship routine, these feathers, which are the same color as the collar, are turned forward and held stiffly in position. Exactly how this is accomplished is not known, but it must involve precise control of minute muscles within the skin that surround the feather follicles. So even a passive visual inspection of color by the female involves active neuromuscular control on the part of the male. The contractions of these tiny skin muscles might even be enhanced by androgens.

Male manakins are unique in erecting this lizard-like dewlap display, but they are not unique in using colorful feathers under the beak as a signal. Think of the gorgets of many hummingbirds, that remarkable display of iridescent color that serves to woo mates and defend territories or sources of nectar. Or consider white-throated sparrows (*Zonotrichia albicollis*), North American birds with dull, slightly streaked throats or with dazzling bright white throats. This and other genetically based morphological differences contribute to an assortative mating system whereby male *and* female birds with fully white throats mate with tan-throated birds of the opposite sex. Throat color is important in this species. Think, too, of the purple-throated fruitcrow (*Querula purpurata*), a largish cotingid relative of manakins that occupies the same Panamanian rainforests as golden-collared manakins. Although the male fruitcrow's overall plumage is a dull blackish-brown, his throat is embellished with an iridescent purple gorget of feathers that can be fanned sideways by a slow twisting of the head. This movement, together with the male's tremulous, mournful call (which gives it its generic name), serves to attract a mate for copulations. In each of these birds, whether hummingbird, sparrow, or fruitcrow, the female must face the male if she is to see the brilliance or pattern of the color signal. Thus, it is a relatively short-distance, intimate effect. The same is presumably true of the beard of *Manacus* manakins.

As we piece all of this biology together, what we see is that a good deal of the male manakin's genetics, morphology, and behavior is at work to gain the attention of the females. Some of the male's signals, rollsnaps and wingsnaps, travel a relatively long distance through dense forest; some of the signals, the color of the beard and the skill of the dance, must be witnessed up close.

By virtue of her very fast visual system, does the female see all of the special neuromuscular skill required to jump, not fly, from sapling to sapling? Does the single midair wingsnap tell the female, in essence, "Hey, look, I don't need my wings to make it from here to there; I can use them for something else, like making crazy snapping sounds"? Is the female's auditory system tuned in such a way that the snap stimulates the hair cells in her ear, causing arousal of her sexual desire?

While our data suggest that females do perceive and carefully assess the information contained in the male's signals, more work is needed. We can, however, appreciate that female manakins have exquisite auditory and visual sensory systems, which they use to assess each male's performance. It is because of the male's evolutionary drive to stimulate the female's senses that the marvelous male manakin courtship behavior came to exist. We are lucky the female manakins have made their choices as they have.

FOURTEEN

Evolution of the Backflip

"I can still see, as though I were looking at a picture, a hall in
a house in Düsseldorf, with a group of children gazing up in
amazement at the banisters on the landing above. There a young
man with long, fair hair was performing the most hair-raising
gymnastic exercises, hanging by his arms and swinging backwards
and forwards, from one side to the other. Finally he swung
himself up until he was balancing on his hands, stretched out his
legs and leapt down into the hall below, landing in the midst of
the admiring children. The young man," she finally reveals, "was
Johannes Brahms; the children were the Schumann family."[1]

Halfway around the world from Panama lies the rich tropical island
of Borneo, the third largest island of the planet. Here, in south-
central Kalimantan (Bornean Indonesia), my wife and I visited the
Tanjung Puting National Park and Orangutan Preserve, where we
interacted with semiwild and wild orangutans. These apes are extraordinarily humanlike—or perhaps I should say we are extraordinarily orangutan-like. Once we were walking along a trail and an
adult orangutan came along, perfectly bipedally, going in the oppo-

site direction, and passed us as any human would do on a Boston sidewalk, paying us absolutely no mind—not that I would expect him to look up and say hi. At another point along a trail, we turned a corner and there, just 2 meters away, sat an adult female orangutan, with another sitting 20 meters beyond. Each female had a toddler, just a few months of age. The youngsters were playing together on the ground in between the mothers, rolling around on top of each other, jumping to and fro with big toothy smiles; periodically they'd stop and look at mom, and then go back to playing. Suddenly, one of the mothers moved a bit, as if to signal it was time to go. Her baby ran to her instantly. The other youngster, sadly realizing the game was over, turned and made his way back to his mother—somersaulting the whole way, nine somersaults in a row, before crawling into her waiting arms.

This toddler was a budding gymnast! But why did it somersault? What benefit could be gained by somersaulting instead of walking or running back to mom? It's unlikely that adult orangutans ever somersault, on the ground or in the trees. As kids, we humans somersault; at least, I sure did. But why? Is it a good motor pattern to acquire early on for some reason? If you fall from a tree, is it better to do a somersault than to fall flat? If you are running from a predator and trip, can a somersault help you regain your feet? Is it just a simple act that improves flexibility and coordination, giving the motoric confidence to more safely move about in the trees or on the ground? But why somersaults? Perhaps daddy orangutans find it hard to get upset with junior when he is somersaulting. Perhaps that mom found it to be just as cute as we did, and experienced a surge of the "love" hormone oxytocin with each somersault performed. I'm sure my wife did. Okay, okay . . . so did I. It was frankly adorable.

Just like this little orangutan, male golden-collared manakins are gymnasts. Indeed, males of many species of manakin, as well as males and females of many other bird species, are exceptional acrobats. In the case of the golden-collared manakin, gymnastic displays are not performed to gain food or to escape predation; they are performed to attract females. I believe that females assess the male's strength, agility, and coordination when choosing whether to mate with him. Why individual males perform exactly as they do remains unknown. Perhaps some males are innately physically superior, or have had more opportunities for practice, or a combination of both. These are questions we still seek to understand.

At this point the reader may be wondering why, of all the birds in the world, I chose this one for study. Why have I devoted so much of my life to this small bird? There are really two answers. One is practical. After obtaining my PhD and looking to develop my own independent research directions, I traveled to Panama to visit my friend Greg Adler, now a professor of biology at the University of Wisconsin, Oshkosh. Back then he was known as the "rat man" of Barro Colorado Island, working at the center for tropical forest research introduced at the beginning of this book, where Frank Chapman first studied the golden-collared (then called Gould's) manakin. Greg is an ecologist who studies rodents in the wild. In Panama, he was looking at the ecological and behavioral factors that sculpt populations of spiny rats on the islands that dot the Panama Canal, including Barro Colorado. Greg is an excellent field biologist, and he and I had done some excellent birdwatching (not to mention other escapades) while traveling through Thailand and Colombia. I was keen to visit him in Panama.

One day while Greg was off playing with his rats, I went birding near Pipeline Road. I came across a group of ornithologists who were mistnetting small birds, and one of them, I believe it was Jeff Brawn from the University of Illinois, Urbana-Champaign, showed me a male golden-collared manakin he had just captured. I had seen and heard the birds displaying in the forest nearby, but this close-up look gave the species a new meaning. I also came to realize that these birds were actually quite common in Panama. This was surprising, as I had simply assumed that an animal that did something so unique and elaborate must be uncommon or even rare. If they are common, I thought, they can be studied. Moreover, the Smithsonian Tropical Research Institute had all of the logistical support I needed.

More important than the practical ability to study this manakin was the fact that this bird is an athlete. Many animals, of course, are exceptional athletes, running or flying or swimming for great distances, or for short distances at great speed; climbing up and down vertical trees or cliffs or swinging elegantly at great heights through dense forest. Some species effortlessly carry around heavy horns or tusks, or dance about wildly to attract a mate. But the golden-collared manakin is not just any kind of athlete. He is a gymnast.

You see, at one time in my life I, too, was a gymnast. When I was about five, my parents, perhaps having noticed that I spent a lot of time doing somersaults, enrolled me in gymnastics classes. These I took to, much like I took to fishing. Gymnastics was natural, easy, and fun for me. I did tumbling, which involved running fast down a long, six-foot-wide cushioned matt before doing a cartwheel, followed by back handsprings or backflips or even backflips with a half twist. I would do floor exercises that added dance or strength moves to the tumbling, all done diagonally across the square gym floor. I did trampoline. (Here, the other boys and I would test each other by doing as many backflips as possible before we'd get so dizzy we'd start to fall off . . . I recall I could do up to forty in a row.) I was especially good at standing or walking on my hands, and would show off by walking the entire perimeter of the gym on my hands. I once walked the length of a football field on my hands. My parents took me to gymnastics tournaments all over the state of Texas where I won some bronze, silver, and gold medals.

When I outgrew gymnastics I turned to springboard diving, 1 and 3 meter, both in high school and then competitively throughout college. Gymnastics and diving involve elaborate and exceptionally well-controlled body movements and strength. My best dive was a back one-and-one-half flip with two and a half twists from the 3-meter board. Nowadays, this dive is considered pretty easy.

To this day, when I see a gymnast produce an exquisite performance, like Simone Biles in the Olympics a few years back, I tear up a bit with a wave of emotion and memories of my own. No wonder I became fascinated by male golden-collared manakins.

But my career in gymnastics didn't come about merely because I enjoyed throwing myself around the backyard, which was noticed by my parents. I was born jaundiced and with an overly curved spine. As a fetus, I possessed the Rh antigen (I was Rh+), which my mother lacked (she was Rh–), and so her immune system went after me; I was delivered by C-section prematurely to limit the damage. My parents must therefore have been relieved to see me as a small boy doing somersaults, a sign of recovery from my fetal and neonatal trauma. So they signed me up for gymnastics classes.

Most kids do all sorts of acrobatics just for fun, on jungle gyms, in trees, at the swimming pool. Many children pursue gymnastics or other solo acrobatic sports under their own motivation or with a little urging from parents. There is some evidence that participation in sports like this can increase bone mass. There is also evidence that building muscles early on can have a lasting influence on muscle size. Epigenetic changes in skeletal muscles, bones, the cardiovascular system, and the central nervous system, all established developmentally, no doubt prepare us for all the motor functions of later life. Think of the confidence we gain by unconsciously knowing how our body moves; we coordinate the movements of so many different neuromuscular systems, inhibiting some, activating others, and doing so impeccably and sometimes with great speed. Undoubtedly, the motor benefits of early acrobatics are crucial for survival into adulthood, a core feature of natural selection.

Humans start somersaulting spontaneously at a very young age. We have anecdotal evidence that male manakins start performing their courtship acrobatics at a young age as well, perhaps just a few months old, but do so weakly and slowly for short periods of time, and usually alone. How does the performance of male manakins change over their lifetime? Do they "improve" with practice? Presumably, males have many of the courtship circuits functional at a young age, and testosterone increases their motivation to practice while also improving and promoting the capabilities of the nerves and muscles to help their practice make perfect.

In golden-collared manakins it is the males that do most of the acrobatics. If tumbling about at a young age gives an individual an evolutionary leg-up, then it raises the question as to whether females of other species perform gymnastics and how their brains are programmed to predispose them to acrobatic behavior.

We have few direct answers, but studies of macaques suggest that hormones do exert some influence at least on the *intensity* of physical play behavior. When very young, macaques tumble about playfully, with that play becoming a bit rougher in males. This all occurs before they reach puberty at three to four years of age. These juvenile males have little to no testosterone, so the behavior seems to be hormone-independent at these ages. The fact that females perform less rough-and-tumble and pursuit play is likely due to hormonal differences during the *fetal* stage of development. Presumably, the neural circuits underlying the motivation to play rough are organized by testosterone; although the circuits are present in the female fetus, they die or decay due to lack of testosterone, whereas in males they are induced to grow in the presence of a fetal surge in testosterone. Fetal females exposed to a small amount of testosterone will play and tumble about as juveniles a bit more roughly than those never exposed to testosterone.

Why would such play behavior be enhanced in males as compared to females? Presumably, the social and motor skills that are used during these forms of play adapt males to a more physical, aggressive life than what females face. Do females prefer to mate with such males? Certainly female primates of some species, including macaques, are aware of the intense physicality of males and avoid them except when ready to copulate.

It is hard not to wonder how neural, hormonal, and muscular systems are organized and activated in humans. Girls and boys both become gymnasts. Does gymnastics have any bearing on human sexual selection, as seems to be the case in manakins? Over the hundreds of thousands of years of human evolution, have motor skills improved because, consciously or unconsciously, choice of a mate is driven by an assessment of such skills, with some talent acquired during early development? Of course, this is a difficult question to answer. Yet, we are all motivated to be physically active, to use our bodies in as many ways as possible, perhaps to stretch the limits with respect to speed, motion, posture, strength, stamina. This motivation is ingrained in our psyche. There are circuits in our brains that are activated when young and drive us to perform such feats. Our cultures capture this drive and form from it all sorts of physically challenging acts. In many cases we build arenas, some especially large, so we can watch other humans show off their physical capabilities.

This athletic drive is a core feature of male manakins. Manakins and humans are much alike. Male golden-collared manakins perform backflips with a half twist. They jump across clearings and wingsnap in midair. They turn, midair, at the last minute to land perfectly. They extend their beards. They remain awkwardly motionless on a perch before jumping and snapping again. And this behavior has meaning to the females.

We have tried to peer into the mind of the adult female manakin by asking which males receive more copulations, but this likely depends on numerous factors, including the quantity and probably quality of the male's acrobatic performance. Perhaps someday we will be able to scan the brains of alert female manakins while they watch males perform, to see which brain areas light up and whether those areas light up more or less based on the specifics of the male's performance, including visual versus auditory stimulation—plumage colors and physical speed, as well as wingsnaps, rollsnaps, and chee-poos. Might a brain scan of a woman watching the men's floor exercise resemble that of a female manakin watching a courting male? Would the same brain areas light up in males watching a female floor exercise? Might the sex of the observer and performer matter not at all in activating the brain? Is it for these reasons that many of us somersault as children?

And what of androgens? To what degree do our neuromuscular systems express receptors for androgens that, at least in males, promote physical performance? Do males who are better in one way or another possess more androgen receptors in their muscles, neurons, and hearts? And what of girls and women? They have some androgens too, and they accomplish remarkable athletic achievements. How are androgen-dependence and androgen-independence balanced in males and in females to shape the exceptional physical performance capabilities of *Homo sapiens?* These are difficult questions to study in humans and remain areas of much debate.

There is much to learn about all of these questions, both in humans and in other species: from the expression of the earliest genes that drive the formation of the reproductive organs and muscles and nerves, to the role of hormones in powering the motivation and physical capability to engage in athletic performance. And what of the psychological strength that comes with practice and the rewards of accomplishment? In all of this, golden-collared manakins are teaching us a great deal. Certainly study of the athletic prowess of other species will teach us even more. We just need to tumble out into nature and discover what it has to offer.

FIFTEEN

What Lies Ahead for Manakins

All athletic accomplishment, including, for male manakins, the behaviors involved in gaining a mate, requires hard work and a "never
quit" attitude. Science is like that as well, with curiosity serving to
motivate action. All of the manakin science described in this book
represents countless hours of devotion by scientists wishing to discover just how things work. For many biologists, that means understanding how organisms, or their parts, function. For naturalists, it
means spending weeks, if not years, in the wild observing organisms
and what they do. It is from just such scientific observation and ex-
perimentation that we know how DNA leads to RNA and then to
proteins that are the basis of an unimaginable number of processes
that enable cell function; how neurons fire and interconnect; how
muscles contract; why hormones are secreted to bind to receptors
and change the properties of cells and tissues. All of these pieces
come together to create life, including the golden-collared manakin.
These processes have also occurred within human brains such as
my own, allowing me to experience the wonders of manakins and
be inspired to devote my life to biological and natural discovery.

I am, of course, not alone in my love for and curiosity about manakins. Through the stories I have told here, you have become acquainted with many of my students and postdocs, most of whom are now successful independent scientists. You have met a few of my colleagues, Dave McDonald, Mike Braun, and Rick Prum among them, who have documented the behavior of several manakin species over the course of many years. There are many others, however, who have gone unmentioned but who have contributed equally to our appreciation of this very special avian group. As you are reading this, many of these individuals are in the field observing manakins; in a lab examining manakin specimens, tissues, RNA or DNA; at their computers analyzing data or writing research papers; or preparing grant proposals to fund the many manakin-related ideas jumping around in their brains.

By the time this book finds its way into your hands, new discoveries will have been made that will expand upon what is presented here. Some of that work may even contradict what is described. That's science. Here are just a few recent advances in golden-collared manakin biology.

Recently, Leo Fusani's research group in Austria developed and utilized a three-dimensional camera array to better follow male and female manakin movements during courtship. This is a very sophisticated setup, requiring computerized synchronization of several two-dimensional camera captures, which are then reconstructed into three dimensions. One observation stands out from their analysis. When a male golden-collared manakin jumps from perch to perch during a courtship display, his acceleration upon takeoff is, on average, 98.63 m/s^2, a leaping force of 9.6 times his body weight. By comparison, the takeoff force of a pigeon, a starling, and a zebra finch is, respectively, 1.7, 2.6, and 4.9 times their body weight. Thus the male manakin exceeds by a factor of two or more the takeoff velocities of other birds. Presumably, this enhanced jumping capability is produced by their enlarged and androgen-sensitive leg muscles.

Another group of scientists, led by James Pease at Wake Forest University, has used powerful bioinformatic techniques to explore the many attributes that make manakin muscles special, particularly the androgen-sensitive and superfast SH muscle that produces wing- and rollsnaps. A plethora of biochemical processes are brought together in this muscle to allow it to repeatedly contract so rapidly. These traits did not all occur at once, but evolved independently over time. In asking just how such a remarkably complex behavior arose, one could consider other manakin species, or related groups of birds, some that perform minimal physical courtship displays and others with highly elaborate performances involving rapidly contracting and relaxing muscles. With the evolution of more elaborate displays, the muscle's proteins became a little better at contracting, at releasing and re-accumulating calcium, and at providing the energy to sustain those contractions. Many of the traits that are required for wingsnaps had begun to evolve uniquely in manakins that did not wingsnap, presumably to accomplish other behavioral functions (like escaping predators). However, when these traits became sensitive to androgenic control, and females liked them, they evolved further into the forms that underlie the golden-collared manakin courtship that we observe today.

In chapter 13 we heard a bit about the hybrid zone, which Mike Braun's group at the Smithsonian studies (see plates 15 and 16). This region is of interest for a variety of reasons. First, it offers a rewarding set of geographical and ecological circumstances for understanding processes of species evolution. Thus, the golden-collared manakin, the white-collared manakin, and the hybrid lemon-collared manakin all grace the cover of Trevor Price's aptly titled book *Speciation in Birds*. A short distance north of this hybrid zone, in the warm Caribbean Sea, lies the Bocas del Toro archipelago, a vast area of small mangrove-covered islets surrounding a chain of true islands that were isolated from mainland Panama by rising seas many millennia ago. *Manacus* manakins exist on some of these islands, where they appear to have evolved into new forms, being physically larger than their mainland counterparts mere tens of kilometers away (a case of insular gigantism). Thus, here in northwest Panama evolution is working on multiple fronts: at the level of species interactions, at a convergence of ecological zones, and through geographic isolation from conspecifics on islands. In this natural laboratory, Braun and various colleagues are using sophisticated bioinformatic techniques to understand just how evolutionary forces sculpt the newly available genomes and, by extension, the anatomy, physiology, and behavior of golden-collared manakins and their allies. Some evidence already suggests that the courtship dance of males across these zones varies in both speed and duration. It will be exciting to determine just what genes are under selection to produce such changes.

A little-known fact is that the famous fictional spy 007 was named for the ornithologist James Bond, an expert in Caribbean birds and author of the definitive *Birds of the West Indies*. Mr. Bond donated funds to the Smithsonian Institute to support research on Caribbean island biology, and James Bond funds are currently supporting some of this manakin research in Bocas del Toro.

For several years, the National Science Foundation has provided funds to support cross-disciplinary work on diverse species of manakins. Dr. Bette Loiselle of the University of Florida leads this enterprise, which includes biologists, geneticists, and individuals with extensive computer expertise. At the present time, we have complete genome sequences of a number of manakin species from diverse genera and with diverse behavioral routines, social structures, and plumage characteristics. Although much of this work remains as yet unpublished, it is profoundly updating our understanding of how genomes evolve.

Let me just mention that manakins as a group have captured the attention of the lay public, thanks in part to the long-running PBS series *Nature*, where they are featured in the opening title sequence.

The internet has many videos of these amazing dancing birds. Go search for one or a few. You'll be glad you did!

Finally, perhaps it goes without saying, but the future of all manakin species is in jeopardy as climate change and habitat loss transform their ecosystems. It is imperative that we commit ourselves to preserving their world, our world, so that generations to come can witness these remarkable animals. By learning more about manakin biology, it is my hope we will be driven with even more urgency to protect natural space for the golden-collared manakin.

Notes

ONE
Setting the Panamanian Stage

1. Andrea Wulf, *The Invention of Nature: Alexander von Humboldt's New World* (New York: Vintage Books, 2015), 319.

FOUR
The Ways and Means of Wingsnapping

1. Frank M. Chapman, "The courtship of Gould's manakin (*Manacus vitellinus vitellinus*) on Barro Colorado Island, Canal Zone," *American Museum of Natural History Bulletin* 68 (1935): 491.
2. Kimberly S. Bostwick and Richard O. Prum, "High-speed video analysis of wing-snapping in two manakin clades (Pipridae: Aves)," *Journal of Experimental Biology* 206 (2003): 3697, 3694.

FIVE
Male Manakins Are Made to Snap

1. Percy R. Lowe, "The anatomy of Gould's manakin (*Manacus vitellinus*) in relation to its display, with special reference to an undescribed pterylar tract

and the attachments of the flexor carpi ulnaris and flexor sublimis digitorum muscles to the secondary wing-feathers," *Ibis* 84 (1942): 58–59.

2. Lowe, "Anatomy of Gould's manakin," 53.
3. Lowe, "Anatomy of Gould's manakin," 57.

SEVEN

Sexual Selection and Mate Choice

1. Charles Darwin, *The Descent of Man, and Selection in Relation to Sex*, vol. 2 (London: John Murray, 1871), 38.

NINE

Hormones Control Behavior in Manakins

1. Rosalyn Yalow, Autobiographical essay, www.nobelprize.org/prizes/medicine /1977/yalow/biographical.

TEN

Male Manakins Keep Their Gardens Clean

1. Chapman, "Courtship of Gould's manakin," 489.

THIRTEEN

Females Shape Male Manakin Behavior

1. Chapman, "Courtship of Gould's manakin," 504, 508. Quoted with permission.
2. Helen Macdonald, *H Is for Hawk* (New York: Grove Press, 2014), 98.

FOURTEEN

Evolution of the Backflip

1. Eugenie Schumann, quoted in Robert Taylor, *Robert Schumann: His Life and Work* (New York: Universe Books, 1982), 310.

References and Recommended Reading

Preface

Goodall, Jane. *In the Shadow of Man.* Boston and New York: Mariner Books and Houghton Mifflin Harcourt, 1971.

Wilson, Edward O. *Sociobiology: The New Synthesis.* Cambridge, MA: Belknap Press of Harvard University Press, 1975.

ONE
Setting the Panamanian Stage

Chapman, Frank. *My Tropical Air Castle: Nature Studies in Panama.* New York: D. Appleton, 1929.

McCullough, David. *The Path between the Seas: The Creation of the Panama Canal, 1870–1914.* New York: Simon & Schuster, 1977.

Vuilleumier, François. "Dean of American ornithologists: The multiple legacies of Frank M. Chapman of the American Museum of Natural History." *Auk* 122 (2005): 389–402.

Wulf, Andrea. *The Invention of Nature: Alexander von Humboldt's New World.* New York: Vintage Books, 2015.

TWO
Those Exuberant Male Manakins

Angehr, George, Dodge Engleman, and Lorna Engleman. *Where to Find Birds in Panama: A Site Guide for Birders.* Panama City: Sociedad Audubon de Panama, 2006.

Chapman, Frank M. "The courtship of Gould's manakin (*Manacus vitellinus vitellinus*) on Barro Colorado Island, Canal Zone." *American Museum of Natural History Bulletin* 68 (1935): 472–521.

Fusani, Leonida, Juli Barske, Chiara Natali, Guido Chelazzi, and Claudio Ciofi. "Relatedness within and between leks of golden-collared manakin differ between sexes and age classes." *Behavioral Ecology* 29 (2018): 1390–1401.

Höglund, Jacob, and Rauno Alatalo. *Leks.* Princeton, NJ: Princeton University Press, 1995.

Johnsgard, Paul. *Arena Birds.* Washington, DC: Smithsonian Institute, 1994.

Lill, Alan. "Sexual behavior of the lek-forming white-bearded manakin (*Manacus manacus trinitatis* Hartert)." *Zeitschrift für Tierpsychologie* 36 (1974): 1–36.

Ridgely, Robert, and John Gwynne. *A Guide to the Birds of Panama, with Costa Rica, Nicaragua, and Honduras.* Princeton, NJ: Princeton University Press, 1992.

Snow, David W. "A field study of the black-and-white manakin, *Manacus manacus*, in Trinidad." *Zoologica* 57 (1962): 65–104.

THREE
Some Ornithology Basics

Brumm, Henrik, and Peter J. B. Slater. "Animals can vary signal amplitude with receiver distance: Evidence from zebra finch song." *Animal Behaviour* 72 (2006): 699–705.

Elemans, Coen P. H., Andrew F. Mead, Lawrence C. Rome, and Franz Goller. "Superfast muscles control song production in songbirds." *PLOS ONE* 2008 Jul 9; 3(7): e2581. doi:10.1371/journal.pone.0002581. PMID: 18612467.

Gill, Frank. *Ornithology.* 3rd ed. New York: W. H. Freeman, 2007.

Goller, Franz, and Roderick Suthers. "Role of syringeal muscles in controlling the phonology of bird song." *Journal of Neurophysiology* 76 (1996): 287–300.

Johnsgard, Paul. *Handbook of Waterfowl Behavior.* Ithaca, NY: Comstock Publishing Associates, Cornell University Press, 1965.

Kroodsma, Donald. *The Singing Life of Birds: The Art and Science of Listening to Birdsong.* Boston: Houghton Mifflin Harcourt, 2005.

Leite, Rafael N., Rebecca T. Kimball, Edward L. Braun, Elizabeth P. Derryberry, Peter A. Hosner, Graham E. Derryberry, Marina Anciaes et al. "Phylogenomics of manakins (Aves: Pipridae) using alternative locus filtering strategies based on informativeness." *Molecular Phylogenetics and Evolution* 2021 Feb; 155:107013. doi:10.1016/j.ympev.2020.107013. PMID: 33217578.

Liu, Wan-chun, Kazuhiro Wada, Erich D. Jarvis, and Fernando Nottebohm. "Rudimentary substrates for vocal learning in a suboscine." *Nature Communication* 13 (2013): 1–12.

Marler, Peter, and Hans Slabbekoorn. *Nature's Music: The Science of Birdsong.* San Diego: Elsevier Academic Press, 2004.

Nottebohm, Fernando, Tegner M. Stokes, and Christiana M. Leonard. "Central control of song in the canary, *Serinus canarius.*" *Journal of Comparative Neurology* 165 (1976): 457–486.

Oliveros, Carl H., Daniel J. Field, Daniel T. Ksepka, F. Keith Barker, Alexandre Aleixo, Michael J. Andersen, Per Alström et al. "Earth history and the passerine superradiation." *Proceedings of the National Academy of Sciences* 116 (2019): 7916–7925.

Reiner, Anton, David J. Perkel, Laura L. Bruce, Ann B. Butler, András Csillag, Wayne Kuenzel, Loreta Medina et al. "Revised nomenclature for avian telencephalon and some related brainstem nuclei." *Journal of Comparative Neurology* 473 (2004): 377–414.

Saldanha, Colin J., J. Douglas Schultz, Sarah E. London, and Barney A. Schlinger. "Telencephalic aromatase, but not a song system, in a sub-oscine passerine, the golden collared manakin (*Manacus vitellinus*)." *Brain, Behavior, and Evolution* 56 (2000): 29–37.

Schmidt, Marc F., and J. Martin Wild. "The respiratory-vocal system of songbirds: Anatomy, physiology, and neural control." *Progress in Brain Research* 212 (2014): 297–335.

Suthers, Roderick A., Franz Goller, and Carolyn L. Pytte. "The neuromuscular control of birdsong." *Philosophical Transactions of the Royal Society B: Biological Sciences* 354 (1999): 927–939.

Suthers, Roderick A., and Daniel Margoliash. "Motor control of birdsong." *Current Opinions in Neurobiology* 12 (2002): 684–690.

Thorpe, William. *Bird-Song: The Biology of Vocal Communication and Expression in Birds.* Cambridge: Cambridge University Press, 1961.

FOUR

The Ways and Means of Wingsnapping

Bodony, Daniel J., Lainy Day, Anthony R. Friscia, Leonida Fusani, Aharon Karon, George W. Swenson Jr, Martin Wikelski, and Barney A. Schlinger. "Determination of the wingsnap sonation mechanism of the golden-collared manakin (*Manacus vitellinus*)." *Journal of Experimental Biology* 219 (2016): 1524–1534.

Bostwick, Kimberly S., and Richard O. Prum. "High-speed video analysis of wing-snapping in two manakin clades (Pipridae: Aves)." *Journal of Experimental Biology* 206 (2003): 3693–3706.

Ventocilla, Jorge, and Kurt Dillon. *Gamboa: A Guide to Its Natural and Cultural Heritage*. Panama City: Smithsonian Tropical Research Institute, 2010.

FIVE

Male Manakins Are Made to Snap

Dial, Kenneth P. "Activity patterns of the wing muscles of the pigeon (*Columbia livia*) during different modes of flight." *Journal of Experimental Zoology* 262 (1992): 357–373.

Friscia, Anthony, Gloria D. Sanin, Willow R. Lindsay, Lainy B. Day, Barney A. Schlinger, Josh Tan, and Matthew J. Fuxjager. "Adaptive evolution of the avian forearm skeleton to support acrobatic display behavior." *Journal of Morphology* 277 (2016): 766–775.

Fuxjager, Matthew J., Leonida Fusani, Franz Goller, Lisa Trost, Andries Ter Maat, Manfred Gahr, Ioana Chiver, R. Miller Ligon IV, Jennifer Chew, and Barney A. Schlinger. "Neuromuscular mechanisms of an elaborate wing display in the golden-collared manakin (*Manacus vitellinus*)." *Journal of Experimental Biology* 220 (2017): 4681–4688.

Fuxjager, Matthew J., Franz Goller, Annika Dirkse, Gloria D. Sanin, and Sarah Garcia. "Select forelimb muscles have evolved superfast contractile speed to support acrobatic social displays." *Elife* 5 (2016): e13544. https://doi.org/10.7554/eLife.13544.

Lowe, Percy R. "The anatomy of Gould's manakin (*Manacus vitellinus*) in relation to its display, with special reference to an undescribed pterylar tract and the attachments of the flexor carpi ulnaris and flexor sublimis digitorum muscles to the secondary wing-feathers." *Ibis* 84 (1942): 50–83.

McDonald, David B. "Correlates of male mating success in a lekking bird with male-male cooperation." *Animal Behavior* 37 (1989): 1007–1022.

McDonald, David B., and Wayne K. Potts. "Cooperative display and relatedness among males in a lek-mating bird." *Science* 266 (1994): 1030–1032.

Schultz, J. Douglas, Fritz Hertel, Monica Bausch, and Barney A. Schlinger. "Adaptations for rapid and forceful contraction in wing muscles of the male golden-collared manakin: Sex and species comparisons." *Journal of Comparative Physiology A* 187 (2001): 677–684.

Uy, J. Albert C., and John A. Endler. "Modification of the visual background increases the conspicuousness of golden-collared manakin displays." *Behavioral Ecology* 15 (2004): 1003–1010.

SIX
Research in Field and Laboratory

"Guidelines for the treatment of animals in behavioural research and teaching." *Animal Behaviour* 135 (2018): i–x.

SEVEN
Sexual Selection and Mate Choice

Barske, Juli, Mansoureh Eghbali, Saritha Kosarussavadi, Eric Choi, and Barney A. Schlinger. "The heart of an acrobatic bird." *Comparative Biochemistry and Physiology A* 228 (2019): 9–17.

Barske, Juli, Leonida Fusani, Martin Wikelski, Ni Feng, M. Santos, Barney A. Schlinger. "Energetics of the acrobatic courtship in male golden-collared manakins (*Manacus vitellinus*)." *Proceedings of the Royal Society B: Biological Sciences* 281 (2013): 0132482. https://doi.org/10.1098/rspb.2013.2482.

Barske, Juli, Barney A. Schlinger, Martin Wikelski, and Leonida Fusani. "Female choice for male motor skills." *Proceedings of the Royal Society B: Biological Sciences* 278 (2011): 3523–3528.

Darwin, Charles. *The Descent of Man, and Selection in Relation to Sex.* London: John Murray, 1871.

Darwin, Charles. *On the Origins of Species by Means of Natural Selection.* London: John Murray, 1859.

Fusani, Leonida, M. Giordano, Lainy Day, and Barney A. Schlinger. "High-speed video analysis reveals individual differences in the courtship display of male golden-collared manakins." *Ethology* 113 (2007): 964–972.

Fuxjager, Matthew J., and Barney A. Schlinger. "Perspectives on the evolution of animal dancing: A case study of manakins." *Current Opinions in Behavioral Science* 6 (2015): 7–12.

Huxley, Julian S. "Darwin's theory of sexual selection and the data subsumed by it, in the light of recent research." *American Naturalist* 72 (1938): 416–433.

Prum, R. O. "Aesthetic evolution by mate choice: Darwin's really dangerous idea." *Philosophical Transactions of the Royal Society B: Biological Sciences* 367 (2012): 2253–2265.

Rosenthal, Gil. *Mate Choice: The Evolution of Sexual Decision Making from Microbes to Humans.* Princeton, NJ: Princeton University Press, 2017.

EIGHT

Behavioral Endocrinology and Vertebrate Sex Differences

Adkins-Regan, Elizabeth. "Hormonal organization and activation: Evolutionary implications and questions." *General and Comparative Endocrinology* 176 (2012): 279–285.

Agate, Robert J., William Grisham, Juli Wade, Suzanne Mann, John Wingfield, Carolyn Schanen, Aarno Palotie, and Arthur P. Arnold. "Neural, not gonadal, origin of brain sex differences in a gynandromorphic finch." *Proceedings of the National Academy of Sciences* 100 (2003): 4873–4878.

Angier, Natalie. *Woman: An Intimate Geography.* Boston: Houghton Mifflin Harcourt, 1999.

Balthazart, Jacques, Arthur P. Arnold, and Elizabeth Adkins-Regan. "Sexual differentiation of brain and behavior in birds." In D. Pfaff and M. Joels, eds., *Hormones, Brain, and Behavior,* 3rd ed., 185–224. New York: Academic Press, 2017.

Chue, Justin, and Craig A. Smith. "Sex determination and sexual differentiation in the avian model." *FEBS Journal* 278 (2011): 1027–1034.

Hines, Mellissa. *Brain Gender.* Oxford: Oxford University Press, 2004.

McCarthy, Margaret M., and Arthur P. Arnold. "Reframing sexual differentiation of the brain." *Nature Neuroscience* 14 (2011): 677–683.

Nelson, Randy, and Lance Kriegsfeld. *An Introduction to Behavioral Endocrinology.* 5th ed. Sunderland, MA: Sinauer Associates, 2016.

Phoenix, Charles H., Robert W. Goy, Arnold A. Gerall, and William C. Young. "Organizational action of prenatally administered testosterone proprionate on the tissues mediating behavior in the female guinea pig." *Endocrinology* 65 (1959): 369–382.

NINE

Hormones Control Behavior in Manakins

Bayliss, William M., and Ernest H. Starling. "The mechanism of pancreatic secretion." *Journal of Physiology* 28 (1902): 325–353.

Berthold, Arnold Adolph. "Transplantation der Hoden." *Archiv fur Anatomie and Physiologie* 16 (1849): 42–46.

Chiver, Ioana, and Barney A. Schlinger. "Sex differences in androgen activation of complex courtship behavior." *Animal Behavior* 124 (2017): 109–117.

Chiver, Ioana, and Barney A. Schlinger. "Sex-specific effects of testosterone on vocal output in a tropical sub-oscine bird." *Animal Behavior* 148 (2019): 105–112.

Day, Lainy B., Leonida Fusani, Estafania Hernandez, Timothy J. Billo, Kimberly S. Sheldon, Petra M. Wise, and Barney A. Schlinger. "Testosterone and its effects on courtship in golden-collared manakins: Seasonal, sex, and age differences." *Hormones and Behavior* 51 (2007): 62–68.

Day, Lainy B., Jennifer T. McBroom, and Barney A. Schlinger. "Testosterone increases display behaviors but does not stimulate growth of adult plumage in male golden-collared manakins." *Hormones and Behavior* 49 (2006): 223–232.

Heimovics, Sarah A., and Lauren V. Riters. "ZENK labeling within social behavior brain regions reveals breeding context-dependent patterns of neural activity associated with song in male European starlings (*Sturnus vulgaris*)." *Behavioral Brain Research* 176 (2007): 333–343.

Nottebohm, Fernando. "Testosterone triggers growth of brain vocal control nuclei in adult female canaries." *Brain Research* 189 (1980): 429–436.

Riters, Lauren V., and Gregory F. Ball. "Lesions to the medial preoptic area affect singing in the male European starling (*Sturnus vulgaris*)." *Hormones and Behavior* 36 (1999): 276–286.

Schlinger, Barney A., Juli Barske, Lainy Day, Leonida Fusani, and Matthew J. Fuxjager. "Hormones and the neuromuscular control of courtship in the golden-collared manakin (*Manacus vitellinus*)." *Frontiers in Neuroendocrinology* 34 (2013): 143–156.

Schlinger, Barney A., and Ioana Chiver. "Behavioral sex differences and hormonal control in a bird with an elaborate courtship display." *Integrative and Comparative Biology* 61 (2021): 1319–1328.

Wikelski, Martin, Michaela Hau, W. Douglas Robinson, and John C. Wingfield. "Reproductive seasonality of seven neotropical passerine species." *Condor* 105 (2003): 683–695.

Wingfield, John C., and Donald S. Farner. "The determination of five steroids in avian plasma by radioimmunoassay and competitive protein-binding." *Steroids* 26 (1975): 311–327.

Wingfield, John C., and Donald S. Farner. "Endocrinology of reproduction in wild species." In D. S. Farner, J. King, and K. C. Parkes, eds., *Avian Biology* 9:163–327. New York: Academic Press, 1993.

TEN

Male Manakins Keep Their Gardens Clean

Ackerman, Jennifer. 2016. *The Genius of Birds*. London: Penguin Books, 2016.

Chiver, Ioana, and Barney A. Schlinger. "Clearing up the court: Sex and the endocrine basis of display-court manipulation." *Animal Behavior* 131 (2017): 115–121.

Clayton, Nicola, and Anthony Dickinson. "Episodic-like memory during cache recovery by scrub jays." *Nature* 395 (1998): 272–274.

Coccon, Francesca, Barney A. Schlinger, and Leonida Fusani. "Male manakins do not adapt their courtship display to spatial alteration of their court." *Ibis* 154 (2012): 173–176.

Craighead, J. J., and F. C. Craighead. *Hawks, Owls, and Wildlife.* New York: Dover, 1969.

Emery, Nathan J. "Cognitive ornithology: The evolution of avian intelligence." *Philosophical Transactions of the Royal Society B: Biological Sciences* 361 (2006): 23–43.

Endler, John A., and Lainy B. Day. 2006. "Ornament colour selection, visual contrast, and the shape of colour preference functions in great bowerbirds, *Chlamydera nuchalis.*" *Animal Behavior* 72 (2006): 1405–1416.

Endler, John A., Lorna C. Endler, and Natalie R. Doerr. "Great bowerbirds create theaters with forced perspective when seen by their audience." *Current Biology* 20 (2010): 1679–1684.

Endler, John A., David A. Westcott, Jonah R. Madden, and Tim Robson. "Animal visual systems and the evolution of color patterns: Sensory processing illuminates signal evolution." *Evolution* 59 (2005): 1795–1818.

Marzluff, John, and Tony Angell. *Gifts of the Crow: How Perception, Emotion, and Thought Allow Smart Birds to Behave like Humans.* New York: Atria Books, 2012.

Strycker, Noah. *The Thing with Feathers: The Surprising Lives of Birds and What They Reveal about Being Human.* New York: Riverhead Books, 2014.

ELEVEN

Male Manakins Are Rich with Androgen Targets

Feng, Ni., Amnon Katz, Lainy Day, Juli Barske, and Barney A. Schlinger. "Limb muscles are androgen targets in an acrobatic tropical bird." *Endocrinology* 151 (2010): 1042–1049

Fusani, Leonida, Juli Barske, Lainy B. Day, Matthew J. Fuxjager, and Barney A. Schlinger. "Sexual selection for male acrobatics in golden-collared manakins: Female choice for neuromuscular skills." *Neuroscience and Biobehavioral Reviews* 46 (2014): 534–546.

Fusani, Leonida, Zoe Donaldson, Sarah E. London, Anahid Mirzatoni, Matthew J. Fuxjager, and Barney A. Schlinger. "Androgen receptor expression in brain of a sub-oscine bird with an elaborate courtship display." *Neuroscience Letters* 578 (2014): 61–65.

Fuxjager, Matthew J., Joy Eaton, Willow R. Lindsay, Lucie H. Salwiczek, Michelle A. Rensel, Juli Barske, Lainy B. Day, and Barney A. Schlinger. "Evolutionary patterns of adaptive acrobatics and physical performance predict expression profiles of androgen receptor—but not estrogen receptor—in the forelimb musculature." *Functional Ecology* 29 (2015): 1197–1208.

Fuxjager, Matthew J., John B. Heston, and Barney A. Schlinger. "Peripheral androgen action helps modulate vocal production in a suboscine passerine." *The Auk: Ornithological Advances* 131 (2014): 327–334.

Fuxjager, Matthew J., Kristy Longpre, Jennifer Chew, Leonida Fusani, and Barney A. Schlinger. "Peripheral androgen receptors sustain the acrobatics and fine motor skill of elaborate male courtship." *Endocrinology* 154 (2013): 3168–3177.

Fuxjager, Matthew J., J. Douglas Schultz, Juli Barske, Ni Feng, Leonida Fusani, Anahid Mirzatoni, Lainy B. Day, Michaela Hau, and Barney A. Schlinger. "Spinal motor and sensory neurons are androgen targets in an acrobatic bird." *Endocrinology* 153 (2012): 3780–3791.

Fuxjager, Matthew J., Eric R. Schuppe, John Hoang, Jennifer Chew, Mital Shah, and Barney A. Schlinger. "Expression of 5α- and 5β-reductase in spinal cord and muscle of birds with different courtship repertoires." *Frontiers in Zoology* 13 (2016): 25. doi: 10.1186/s12983-016-0156-y.

Prum, Richard O. "Phylogenetic analysis of the evolution of display behaviour in the neotropical manakins (Aves: Pipridae)." *Ethology* 84 (1990): 202–231.

Schlinger, Barney A. "Steroids in the avian brain: Heterogeneity across space and time." *Journal of Ornithology* 156 (2015): 419–424.

Schultz, J. Douglas, and Barney A. Schlinger. "Widespread accumulation of 3H-testosterone in the spinal cord of a wild bird with an elaborate courtship display." *Proceedings of the National Academy of Sciences* 96 (1999): 10432–10436.

Sewall, Kendra B. "Androgen receptor expression could contribute to the honesty of a sexual signal and be the basis of species differences in courtship displays." *Functional Ecology* 29 (2015): 1111–1113.

TWELVE

Manakin Muscles Are True to the Task

Fuxjager, Matthew J., Juli Barske, Senmi Du, Lainy B. Day, and Barney A. Schlinger. "Androgens regulate gene expression in avian skeletal muscles." *PLOS ONE* 7, no. 12 (2012): e51482. doi:10.1371/journal.pone.0051482.

Fuxjager, Matthew J., Jae-Hyung Lee, Tak-Ming Chan, Jae Hoon Bahn, Jennifer G. Chew, Xinshu Xiao, and Barney A. Schlinger. "Hormones, genes, and athleticism: Effect of androgens on the avian muscular transcriptome." *Molecular Endocrinology* 30 (2016): 254–271.

Fuxjager, Matthew J., Meredith C. Miles, Franz Goller, John Petersen, and Julia

Yancey. "Androgens support male acrobatic courtship behavior by enhancing muscle speed and easing the severity of its trade-off with force." *Endocrinology* 158 (2017): 4038–4046.

Pradhan, Devaleena S., Chunqi Ma, Barney A. Schlinger, Kiran K. Soma, and Marilyn Ramenofsky. "Preparing to migrate: Differential expression of androgen signaling and IGF-1 in muscles of a long-distance migrant." *Journal of Comparative Physiology A.* 205 (2019): 113–123.

THIRTEEN
Females Shape Male Manakin Behavior

Barske, Juli, Barney A. Schlinger, and Leonida Fusani. "The presence of a female influences courtship performance of male manakins." *The Auk: Ornithological Advances* 132 (2015): 594–603.

Brumfield, Robb T., Robert W. Jernigan, David B. McDonald, and Michael J. Braun. "Evolutionary implications of divergent clines in an avian (Manacus: Aves) hybrid zone." *Evolution* 55 (2001): 2070–2087.

Candolin, Ulrika. "The use of multiple cues in mate choice." *Biological Reviews* 78 (2003): 575–595.

Day, Lainy B., Leonida Fusani, Carol Kim, and Barney A. Schlinger. "Sexually dimorphic neural phenotypes in golden-collared manakins (*Manacus vitellinus*)." *Brain, Behavior, and Evolution* 77 (2011): 206–218.

DuVal, Emily H., and Bart Kempenaers. "Sexual selection in a lekking bird: The relative opportunity for selection by female choice and male competition." *Proceedings of the Royal Society B: Biological Sciences* 275 (2008): 1995–2003.

Hau, Michaela, Martin Wikelski, and John C. Wingfeld. 1998. "A neotropical forest bird can measure the slight changes in tropical photoperiod." *Proceedings of the Royal Society B: Biological Sciences* 265 (1998): 89–95.

Macdonald, Helen. *H Is for Hawk*. New York: Grove Press, 2014.

McDonald, David B., Robert P. Clay, Robb T. Brumfield, and Michael J. Braun. "Sexual selection on plumage and behavior in an avian hybrid zone: Experimental tests of male-male interactions." *Evolution* 55 (2001): 1443–1451.

Stein, Adam C., and J. Albert C. Uy. "Plumage brightness predicts male mating success in the lekking golden-collared manakin *Manacus vitellinus*." *Behavioral Ecology* 17 (2006): 41–47.

Uy, J. Albert C., and John A. Endler. "Modification of the visual background increases the conspicuousness of golden-collared manakin displays." *Behavioral Ecology* 15 (2004): 1003–1010.

References and Recommended Reading

Uy, J. Albert C., and Adam C. Stein. "Variable visual habitats may influence the spread of colourful plumage across an avian hybrid zone." *Journal of Evolutionary Biology* 20 (2007): 1847–1858.

FOURTEEN
Evolution of the Backflip

Royte, Elizabeth. *The Tapir's Morning Bath: Solving the Mysteries of the Tropical Rain Forest.* Boston: Houghton Mifflin Harcourt, 2001.

FIFTEEN
What Lies Ahead for Manakins

Janisch, Judith, Elisa Perinot, Leonida Fusani, and Cliodhna Quigley. "Deciphering choreographies of elaborate courtship displays of golden-collared manakins using markerless motion capture." *Ethology* 127 (2021): 550–562.

Pease, James B., Robert J. Driver, David A. de la Cerda, Lainy B. Day, Willow R. Lindsay, Barney A. Schlinger, Eric R. Schuppe, Christopher N. Balakrishnan, and Matthew J. Fuxjager. "Layered evolution of gene expression in 'superfast' muscles for courtship." *Proceedings of the National Academy of Sciences* 119, no. 14 (2022): e2119671119. doi:10.1073/pnas.2119671119.

Price, Trevor. *Speciation in Birds.* Greenwood Village, CO: Roberts & Co., 2007.

Acknowledgments

It has been a real joy to think back over my life to all of the individuals who helped steer me along the circuitous path of becoming the biologist that I am today. First, my deep appreciation to my parents, Norma and Henry, who allowed boy Barney to go off fishing on his own, to live within and ponder about nature. They also bought me my little five-gallon aquarium and the Tasco microscope and chemistry set that occupied my solo boyhood play and thinking for years. Their patience was often tested, as these "toys" periodically produced dried-up fish or snails entangled in the fibers of my bedroom carpet or regularly stained the ceiling with exploded chemicals.

Countless individuals stimulated and shared my passion for animal behavior, birds, and birdwatching, beginning with my animal behavior instructor at Tufts University, Ben Dane, and the undergraduate and graduate students in Ben's weekly Animal Behavior seminar. Every Wednesday evening for several semesters of college, we met to discuss scientific papers, but after some wine and food the conversation invariably devolved into stories about birdwatching and fishing. And for that, I received course credit! Norton Nickerson also stands out as a professor who engendered a love of biology in all of his students. I particularly recall the spring-break trip to the Bahamas, where I banded migrant birds for the first time (and also fished for bonefish while walking the shallows among small sharks and rays and curious royal terns). George Peabody was a fellow student birdwatcher who helped me realize that birdwatchers aren't necessarily nerds. George Gove, a professor at Tufts and many years my senior, became my birdwatching mentor. He also forced me out, spotting scope in hand, to scan for snowy owls or rough-legged hawks on the bitterest of cold Plum Island days while he waited in the heated car. Bobbi Parker, though not a biologist, shared many adventures and tolerated my very early mornings away chasing birds. Robert Berezin encouraged all these pursuits.

After college, I spent a number of months as an intern at the Manomet Bird Observatory on Cape Cod in Massachusetts, where I learned bird-banding techniques and lived in a like-minded community of bird enthusiasts. Similarly, I found a community of biologists while working summers at Parker River National Wildlife Refuge (on the very same Plum Island), doing manual labor and teaching environmental science to high school students. I owe the individuals from those times, both staff and fellow interns, a great deal of thanks.

From graduate school on I have shared my life as a biologist with close friends, leading bird- and whale-watching trips, driving great distances to find new birds, clambering midwinter to the end of the rock jetty at Salisbury Beach in northern Massachusetts in hopes of scaring up Iceland and glaucous gulls or even a rare ivory gull, or trudging through parasite-ridden rainforest swamps or flooded firefly-laden forests of the southeastern United States. These individuals include Greg Adler, Bill Dolphin, Mitch Heindel, Dave Morimoto, and Mark Wourms. As director of the Bernheim Arboretum and Research Forest outside Louisville, Kentucky, Mark also gave me housing as a resident ornithologist at Bernheim to work, in part, on this manuscript.

Much of my success as a scientist has been framed by the very capable mentoring I have received. I am especially indebted to graduate advisors Mike Baum, Gloria Callard, Ian Callard, and Fred Wasserman. I owe special thanks as well to my postdoctoral advisor, and now colleague and close friend, Art Arnold. Many other academic colleagues at UCLA have added wisdom and adventure to my life, for which I am grateful.

Copious thanks go to those who read drafts of individual chapters or the whole book at various stages of development and made recommendations, both small and large, that improved the overall quality. These include Art Arnold, Dan Blumstein, Daniel Bodony, Bill Dolphin, Leo Fusani, Matt Fuxjager, Alan Grinnell, Jeanne Kelley, Colin Saldanha, and Tom Smith.

I owe a special debt of gratitude to my sister Paula, who for many years has encouraged my writing and who pushed me gently forward with this book in particular. Jeanne Kelley and Joe Colichio encouraged my early writing as well. My brother Hank kept me sane by providing me with a steady supply of new classical music.

My research has been funded by the National Science Foundation and by the UCLA Academic Senate. The work would not have been possible without logistical assistance from the staff of the Smithsonian Tropical Research Institute, especially Maria Leone and Raineldo Urriola.

I have been fortunate to have had many exceptional students, both undergraduate and graduate, and postdoctoral fellows working with me in the lab and in the field. They include Juli Barske, Virginie Canoine, Ioana Chiver, Lainy Day, Zoe Donaldson, Joy Eaton, Jenny (Ni) Feng, Leonida Fusani, Matthew Fuxjager, Sarah London, Anahid Mirzatoni, Luke Remage-Healey, Michelle Rensel, Colin Saldanha, Doug Schultz, and Kiran Soma, and I am grateful for their contributions. The same goes for many terrific colleagues across scientific disciplines with whom I've collaborated, including Daniel Bodony, Mike Braun, Tony Friscia, Ted Goslow, Fritz Hertel, and Grace Xiao, and, at the Max Planck Institutes for Ornithology and Behavior, Manfred Gahr, Ebo Gwinner, Michaela Hau, Lisa Trost, and Martin Wikelski.

I owe special thanks to Jean Thomson Black, senior executive editor at Yale University Press, for believing in the project and moving the book along the path to publication. Thanks also to Mai Reitmeyer at the American Museum of Natural History and Russell Johnson at UCLA for assistance with some images reproduced from early publications and to Bill Nelson for preparing several of the illustrations. And thanks to my copyeditor, Anne Canright.

There are no words to express my deep appreciation and love for my wife, Lorie, who for the past twenty years has encouraged me at every step, at home and in the field. Her smile, her joyful personality, and her inner courage have been a mainstay for me. She read or heard every chapter of this book and offered many useful suggestions.

And of course I am grateful for my dogs, Wilson, Gita, and, most recently, Buddha.

I may well have forgotten individuals who have, in one way or another, contributed to making this research and this book come to fruition. I apologize to any I have overlooked.

Index

Pages with illustrations are in italic.